CAN THE LAWS OF

Physics Be Unified?

CAN THE LAWS OF

Physics Be Unified?

PAUL LANGACKER

PRINCETON UNIVERSITY PRESS

PRINCETON AND OXFORD

Copyright © 2017 by Princeton University Press

Published by Princeton University Press, 41 William Street,
Princeton, New Jersey 08540

In the United Kingdom: Princeton University Press, 6 Oxford Street,
Woodstock, Oxfordshire OX20 1TR

press.princeton.edu

Jacket design by Jess Massabrook

All Rights Reserved

Library of Congress Cataloging-in-Publication Data

Names: Langacker, P., author.
Title: Can the laws of physics be unified? / Paul Langacker.
Other titles: Princeton frontiers in physics.
Description: Princeton ; Oxford : Princeton University Press, [2017] |
    Series: Princeton frontiers in physics
Identifiers: LCCN 2016042937 | ISBN 9780691167794
    (hardcover ; alk. paper) | ISBN 0691167796 (hardcover ;
    alk. paper)
Subjects: LCSH: Particles (Nuclear physics) | Standard model
    (Nuclear physics)
Classification: LCC QC793.2 .L35 2017 | DDC 530.14/23—dc23
    LC record available at https://lccn.loc.gov/2016042937

British Library Cataloging-in-Publication Data is available

This book has been composed in Adobe Garamond and
Helvetica Neue LT Std

Printed on acid-free paper ∞

Typeset by Nova Techset Pvt Ltd, Bangalore, India
Printed in the United States of America

1 3 5 7 9 10 8 6 4 2

CONTENTS

PREFACE

I somehow knew even in high school that I wanted to go into particle physics. Perhaps it was because of Leon Lederman's *Scientific American* article on "The Two-Neutrino Experiment" (Lederman 1963), or perhaps it was because of a series of lectures that I attended in Chicago given by Willie Fowler on nuclear synthesis in the Sun. Whatever the reason, I was hooked. When I began graduate school at Berkeley in 1968, much was already known about the properties of the particles and their interactions. Quantum electrodynamics was well established. Experimentation at ever higher-energy cyclotrons and synchrotrons had discovered large numbers of elementary particles and much about their strong and weak interactions. There were promising theoretical models of aspects of particle physics, but little hope of developing a fundamental mathematical description of the strong or weak interactions in the foreseeable future.[1]

Nevertheless, a revolution in our understanding was about to occur. In less than a decade, the standard model

---

[1] One of the seminal papers on the standard model had already been published (Weinberg 1967), but was not widely recognized until some years later.

(more properly called the standard theory) of the strong, weak, and electromagnetic interactions was developed and partially confirmed. It did not come out of nowhere, but rather incorporated and synthesized many of the earlier ideas. Over the next 40 years or so, the standard model (extended to include a third family of particles and neutrino mass) was experimentally verified in exquisite detail, culminating in the discovery of the Higgs boson in 2012. It correctly describes nature, at least to an excellent first approximation, in a mathematically consistent way down to a distance scale smaller than $1/1000^{\text{th}}$ the size of the atomic nucleus.

However, the standard model is incomplete. It is simple and elegant in its basic structure, but extremely complicated in detail. It does not fully unify the known microscopic interactions and has many unexplained parameters that must be taken from experiment. It does not incorporate a quantum theory of gravity; explain the observed dark matter and dark energy of the Universe; or account for the excess of matter over antimatter. There are intriguing theoretical ideas for approaching these issues, which however mainly manifest themselves at shorter distances and higher energies than current experiments.

I have been personally fortunate in that my professional career coincided closely in time with the establishment of the standard model. In this volume, I hope to convey some of the wonder and excitement of particle physics, and in the development, implications, and shortcomings of the standard model, as well as to describe something of the theoretical ideas and experimental prospects for the future. It is my hope that sometime in the next 10, 50, or 100 years, we will successfully develop and (at least) indirectly verify an even greater synthesis, which will truly unify all of the interactions, including quantum gravity; perhaps address the origin of space and time; and provide the framework for understanding the large-scale structure, origin, and ultimate fate of the Universe. We may never fully accomplish this dream, but it is a worthy goal.

A Note to the Reader

This book is written for an undergraduate physics student, a practicing scientist in a related field, or any interested reader familiar with the basic ideas of classical physics, quantum theory, and relativity.

Much of the technology in modern particle physics is rather abstract and technical. In order to present the central concepts, problems, and future possibilities at more than a popular level, I have introduced the rudiments of such topics as relativistic quantum theory, Feynman diagrams, and gauge theories. These will be easier for some readers than others, but I hope that anyone with a midlevel undergraduate background in physics will be able to follow enough of the development to appreciate the remainder.

Chapters 1 through 3 should be easily understandable to readers with the stated background. The most challenging parts are sections 4.1 through 4.3, which introduce field theory, internal symmetries, and Yang-Mills theories. I have tried to present the material in such a way that the essential ideas should be accessible even if the mathematical details are not. Sections 4.4

through 4.7 describe our current understanding of the strong and electroweak interactions, the Higgs boson discovery, and neutrino physics. These utilize some of the formalism, but are generally more qualitative and should hopefully be understandable even for readers who have skimmed over the more technical materials. The remaining chapters deal with the problems of the standard model and possibilities for the future. These should be reasonably straightforward.

A confusing variety of terms and acronyms are used in particle physics and cosmology. Most of those relevant here are defined in the extensive glossary, or can be located through the index.

More detailed discussions of particle physics and the standard model can be found in a number of undergraduate (e.g., Griffiths 2008; Mann 2010; Thomson 2013) and graduate (e.g., Langacker 2010; Tully 2011; Quigg 2013) texts. The bibliography contains a mixture of original papers, technical articles, and pedagogical introductions and reviews. The latter are indicated by a double dagger (††) preceding the title.

CAN THE LAWS OF

Physics Be Unified?

1

THE EPIC QUEST

Curious individuals have speculated about nature and humankind's place in it for all of recorded history. The interests of the ancient Greeks included the structure of matter on the smallest scale and that of the Universe on the largest (see, for example, Weinberg 2015). Some of their ideas were surprisingly similar to our modern understanding. Leucippus and his student Democritus proposed in the fifth century BCE an atomic theory in which matter ultimately consists of indivisible atoms that come in various sizes and shapes, accounting for the myriad materials and their properties that we observe. Aristarchus of Samos later proposed a heliocentric cosmology in which the planets rotated around the Sun and the distant stars were similar in character to the Sun. The technology did not exist to test either of these ideas until millennia later, and in fact there were alternative ideas that were more widely believed. Nevertheless, they illustrate the ingenuity and the craving for understanding of the human mind.

The atomic theory was not completely established until the nineteenth and early twentieth centuries, and the understanding of the structure of the atom and of the quantum-mechanical rules that govern its behavior were not fully worked out until some decades later. These ideas, combined with the parallel understanding of electricity and magnetism and of the kinetic theory, form the physical basis for chemistry, electronics, macroscopic matter in all its forms, and even biology.

By the late 1920s, it was known that the atom consists of a cloud of one or more electrons held in place by electrical forces as they orbit a tiny but very massive nucleus. Furthermore, the dynamics are governed not by the venerable classical mechanics of Isaac Newton, but by the weird quantum mechanics according to which electrons seem to be both particles and waves simultaneously. However, this understanding raised many more questions, such as the details of atomic transitions from one level to another. Similarly, what was the nature of the nucleus? Could the different nuclei somehow be composed of protons and electrons? What were the rules that govern the radioactive decays of nuclei that had been observed in the late nineteenth century? How could quantum mechanics, which governs the very small, be combined with Albert Einstein's relativity, which modifies the notions of space and time for rapidly moving observers or in the presence of matter?

Theoretical and experimental developments in the decades surrounding World War II answered some of these questions while raising others. Quantum theory and special relativity were elegantly combined in the Dirac theory and in the quantization of the electromagnetic field. The former predicted the existence of antimatter, which was subsequently observed. The neutron was discovered, and the basic structure of the nucleus as consisting of protons and neutrons held together by a complicated new *strong interaction* were gradually understood. Similarly, the properties of the *weak interaction*, which is responsible for one form of radioactivity (β *decay*) were gradually worked out, and the apparent non-conservation of energy in β decay was finally understood to result from the non-observation of an almost ghostly *neutrino*. The interactions of high-energy particles from cosmic rays or that were artificially accelerated in cyclotrons and subsequent particle accelerators led to the discovery of additional fundamental particles and of systematic properties of their interactions.

These advances led to the new fields of nuclear physics and then of *elementary particle* (or *high-energy*) physics, which sought to systematically understand the properties of the smallest constituents of nature and their interactions. For some 20 years, there was both confusion and painfully slow progress, involving mathematical difficulties in the theories and a proliferation of particles. Finally, however, what we now call the *standard model* (SM) of the elementary particles and their interactions was completed by the early 1970s. In the next 40 years or so, essentially all of the predictions and ingredients of the SM were experimentally verified, often in great detail, the most recent being the discovery of the *Higgs boson* in 2012. The standard model is a mathematically consistent theory that accounts for essentially all aspects of ordinary matter down to a distance scale of $\mathcal{O}(10^{-16}$ cm$)$.

Despite these successes, the standard model is incomplete: It is very complicated and apparently arbitrary.

The strong, weak, and electromagnetic interactions are very different. Although all are based on the elegant concept of *gauge invariance*, they are not truly unified with each other. A quantum-mechanical description of gravity is not included, although classical general relativity can be grafted on. Furthermore, the SM involves numerous fundamental parameters whose values are not explained. Some of these appear to be fine-tuned to incredibly small (but nonzero) values. There is no explanation of why the electric charges of all particles are integer multiples of $e/3$, where $-e$ is the charge of the electron, nor is there an explanation of the observed excess of matter over antimatter. The neutrinos, initially thought to be massless, are now observed to have nonzero masses much smaller than those of the other fundamental particles. The origin and even nature of these oddball masses are yet to be determined. Finally, the astronomers have determined that the ordinary matter that we are made of and that is described by the SM is only a small fraction of the matter and energy in the Universe. The natures of the dark matter and dark energy are unknown. There is almost certainly a more fundamental desciption of nature that incorporates and extends the SM, generally referred to as *new physics* or *beyond the standard model* (BSM).

There are many ideas for "bottom-up" extensions, with such names as *supersymmetry*, *compositeness*, or *extra space dimensions*, which address some of these issues and which might be manifested in future accelerator and other experiments. "Top-down" ideas, such as *grand unification* or the even more ambitious *superstring* theories could possibly lead to an ultimate unification of the microscopic forces, perhaps including quantum gravity and tackling the origin of space and time. These mainly manifest themselves at incredibly short distance scales that are nearly impossible to directly probe experimentally (with *proton decay* a notable exception), but they might be tested indirectly by their predictions for the low-energy parameters or for new particles or interactions.

There have also been enormous advances since the time of the ancients in our understanding of nature on large scales, including the motions of the Solar System, the composition and energetics of the Sun and stars, of our Galaxy, and of the vast collections of galaxies extending across the fourteen billion light-year radius of the observable Universe.[1] Furthermore, the Universe is expanding and cooling, and can be traced backward to a *big bang* some 14 billion years ago, when it was incredibly hot and dense. Although astronomy and cosmology are not the main thrusts of this volume, they cannot be entirely ignored. The visible parts of stars, galaxies, and other astronomical objects are composed of the same atoms, molecules, nuclei, and elementary particles that we observe in the laboratory, and their dynamics are driven by these particles and their interactions. Even the dark matter is likely due to some still-unobserved elementary particle, while the dark energy may be associated with the ground state (vacuum) energy of some of the fundamental particles. There is even the intriguing suggestion from superstring theory that our observable Universe might

---

[1] The Universe could be much larger, but we can observe only as far as light has traveled since the big bang.

be but a tiny bubble in a vast *multiverse* of regions, each with different laws of physics! The physics of very small distances and astrophysics/cosmology have become inextricably linked.

The atomic theory and the standard model complete two important chapters in the epic quest begun by the ancient Greeks and others to understand the nature in which we live. Parallel chapters in astronomy include the undertanding of the Solar System, the discovery of galaxies, and the expanding Universe/big bang. Although there have been an enormous range of practical applications (especially of atomic physics) and spinoff technologies,[2] the most important aspect for many is simply curiosity about how nature works at her most fundamental level. The combination of new experimental and observational tools, as well as promising theoretical ideas, gives us the chance of even more exciting chapters yet to come on the very small, the very large, and their relation.

---

[2] Particle physics has contributed to many important spinoff technologies, including medical diagnostics and therapies, cryogenics, magnet technology, complex electronics, large-scale distributed computing, and the World Wide Web. Mathematical techniques have found application in other branches of physics. Finally, large experiments and labs have been a remarkable model for international cooperation.

2

THE THREE ERAS

2.1 The Ingredients

The description of any physical system requires three ingredients: (1) What are the basic entities to be described? (2) What are the forces or influences acting on them? (3) What are the rules of the game, e.g., how do the entities respond to those influences? For example, Newtonian gravity involves entities such as the Sun, Moon, Earth, apples, or people. The force is gravity, $\vec{F} = G_N m_1 m_2 \hat{r}/r^2$, and the response is given by Newton's laws, especially $\vec{F} = m\vec{a}$. In general relativity, space-time is added to the list of entities, which is distorted by matter, while point masses respond by following geodesics.

The essence of high-energy (particle) physics is the description of nature at the most fundamental level. At our present level of understanding, the basic entities are elementary particles such as *quarks* (constitutents of the proton and neutron) and *leptons* (e.g., the electron and neutrinos). These appear to be *point-like*, i.e., no evidence has been observed for a nonzero size, or that they are composites of still smaller objects or of a continuous distribution of matter.

The quarks and leptons are acted on by at least five types of *interactions*.[1] These are the strong, which binds the quarks and nucleons together; the electromagnetic, responsible for atomic and molecular binding; the weak, responsible for β decay; the *Higgs-Yukawa*, associated with their mass; and the gravitational, which is mainly important for macroscopic objects. These have very different properties. For example, electromagnetism and gravity are long-range, i.e., the forces between two particles fall off slowly with their separation, while the strong interaction between nucleons becomes insignificant for separations much larger than the size of the nucleus. The properties of the known particles and interactions will be described more fully in chapter 3.

The framework is that of relativistic *quantum field theory*, which is the union of quantum mechanics, special relativity, and the possibility of particle creation or annihilation (such as the reaction $e^+e^- \rightarrow p\bar{p}$ via an intermediate virtual photon). Our understanding of these issues may eventually be supplanted by something more basic, just as Newtonian gravity was superseded by general relativity.

Another issue is often ignored or taken for granted: are the laws of nature absolute? That is, are they uniquely determined, perhaps by some underlying selection principle or by self-consistency, and are they the same everywhere

---

[1] The term *interaction* is more appropriate than *force*, in part because interactions can describe particle creation, annihilation, or transitions from one type of particle to another.

in space-time? That is the implicit assumption of most physicists, and we have not seen any conclusive empirical evidence to the contrary. However, developments in superstring theory and cosmology have caused some to question the absoluteness of physics, as will be described in chapter 5.

2.2 Prehistory

One of the precepts of classical physics is absolute space and time. Space is simply the stage on which events occur, and time keeps track of their sequence. They are the same for all observers, and space obeys the flatness axioms of Euclidean geometry. Another cornerstone is determinism: given an exact knowledge of the initial conditions of a system, one can in principle calculate its future evolution. Both of these precepts were shattered in the early twentieth century. Einstein's special relativity of 1905 showed that space and time and even the sequence of events depend on the motion of the observer, while general relativity (1916) showed that space-time need not be flat.

Similarly, quantum theory replaced determinism by uncertainty and probability. We will focus on the wave mechanics formulation, which describes the motion of a particle of mass m moving in a potential $V(\vec{x}, t)$. Instead of the particle following a deterministic trajectory, it satisfies the Schrödinger equation (1926)

$$\left(-\frac{\hbar^2}{2m}\nabla^2 + V\right)\psi = i\hbar\frac{\partial\psi}{\partial t} = E\psi, \qquad (2.1)$$

where the last form refers to an energy eigenstate solution with definite energy E, for which $\psi(\vec{x}, t) = \psi(\vec{x})e^{-iEt/\hbar}$. For negative energies (bound states), only discrete values of E are allowed, i.e., the system is quantized. The wave function $\psi(\vec{x}, t)$ was later interpreted by Max Born as the probability amplitude: $|\psi(\vec{x}, t)|^2$ is the probability of finding the particle at position $\vec{x}$ at time t. Equation (2.1) gives an accurate description of the spectra of hydrogen and other light atoms (after including spin and the Pauli exclusion principle).

After a digression on units, we will turn to the subsequent development of our understanding of particle and nuclear physics, which I divide into the *Era of Exploration*, the *Standard Model Era*, and the *Beyond the Standard Model Era*.

A Digression: Particle Units

Let us briefly digress on *particle units*, $\hbar = 1, c = 1$, a very convenient compact notation useful in particle physics that will be employed in the following. Setting c (the speed of light in vacuum) to unity implies that all velocities are dimensionless quantities expressed as a fraction of the speed of light. It also implies that distance and time have the same units (e.g., a light-second, or just second, is the distance that light travels in one second), and that mass, energy, and momentum all have the same units. Thus, the usual relation $E^2 = \vec{p}^2 c^2 + m^2 c^4$ between the mass (m), momentum ($\vec{p}$), and energy (E) of a particle becomes $E^2 = \vec{p}^2 + m^2$. Ordinary units can be restored by multiplying a quantity by appropriate powers of $c \sim 3.0 \times 10^{10}$ cm/s, and using the relation 1 eV/$c^2 \sim 1.8 \times 10^{-33}$ g between grams (g) and electron volts (eV), the energy acquired by an electron accelerated through a one volt potential.[2]

The convention $\hbar = 1$ is motivated by the wave-particle duality of quantum mechanics, and in particular by the de Broglie relation $\lambda = 2\pi\hbar/p$ between wavelength and momentum. Thus, in particle units distance can be expressed in units of energy^{-1} and vice versa. Again, ordinary units can be restored by multiplying by powers of $\hbar \sim 6.6 \times 10^{-16}$ eV-s, and $\hbar c \sim 1.97 \times 10^{-5}$ eV-cm.

It is convenient to introduce the *Planck scale*, $M_P = G_N^{-1/2} \sim 10^{28}$ eV, where G_N is the gravitational constant. Its value $\sim 6.7 \times 10^{-8}$cm^3 g^{-1} s^{-2} becomes 6.7×10^{-57} eV^{-2} in particle units. Its significance is that the coefficient of $1/r^2$ in the Newtonian force law is the dimensionless ratio $m_1 m_2/M_P^2$. Gravity becomes strong for masses (or energies in the generalization to general relativity) of $\mathcal{O}(M_P)$, and quantum gravity effects then become important.

Particle units emphasize that in some sense, c and $\hbar$ are not really fundamental quantities, and the conventional values are observable only due to the historical accidents about how such units as g, s, cm, and eV were defined. Only dimensionless ratios such as v/c, the fine structure constant $\alpha = e^2/4\pi$, and ratios of particle masses to each other or to the Planck scale are physically meaningful.

---

[2]Related energy units, such as GeV, are defined in the glossary.

2.3 The Era of Exploration

The era of exploration refers to the development of particle physics prior to the standard mode. The period spans roughly from the quantization of the electromagnetic field[3] by Paul Dirac in 1927 to the development of the *electroweak* $SU(2) \times U(1)$ *model* (late 1960s) or the development of *quantum chromodynamics* (QCD) (early 1970s).

Field quantization refers to the replacement of a quantum wave function or a classical field by an operator that contains creation and annihilation operators. Similar to those of the simple harmonic oscillator, these act on a quantum state to obtain a state involving respectively one more or one less particle or quantum of the field. For each frequency $\omega/2\pi$, direction, and polarization of the electromagnetic field, for example, there can be an integer number of quanta (*photons*), each carrying energy $\hbar\omega$, i.e., ω in particle units. Field quantization allowed a description of atomic transitions involving the emission or absorption of a photon.

The Dirac equation (1928) is a relativistic wave equation for a spin-1/2 particle. Remarkably, this union of special relativity with quantum mechanics predicted the existence of antimatter, i.e., the Dirac equation for the electron had additional solutions corresponding to a positively charged particle. After some confusion that this particle might be the proton, it became clear that it had to have

---

[3] Detailed discussions of the history are given in Pais (1986) and Weinberg (1995).

the same mass as the electron. The existence of this anti-electron, now known as the *positron* (e^+), was confirmed in 1932 by Carl Anderson's observation of positron tracks from cosmic rays in a cloud chamber.

Quantum electrodynamics (QED) was developed experimentally and theoretically in the subsequent two decades or so (see, for example, Schwinger 1958). QED, which involves quantized Dirac and electromagnetic fields, combines classical electrodynamics, quantum mechanics, and special relativity. It explains subtleties such as the Lamb shift in hydrogen and the anomalous magnetic moment of the electron. Early on, QED exhibited seemingly insuperable mathematical difficulties involving infinities encountered in summing over intermediate states in perturbative calculations. By around 1950, however, it was understood that these infinities could be cured by *renormalization* theory, in which physical observables are expressed in terms of the measured values of quantities such as mass and charge rather than the parameters in the original equations of motion.[4] QED is mathematically consistent down to incredibly small distance scales, and has now been tested at the precision of 10^{-8}–10^{-9} in many experiments. It is generally spectacularly successful, although in recent years two anomalies, possibly due to new physics, have emerged. These will be described in chapter 4.

In parallel with the development of QED, the 1930s witnessed significant progress in understanding the structure of the nucleus and the nature of the strong

---

[4]The modern view is that the infinities never really appear, since sums over intermediate states are truncated by, e.g., the Planck scale, $M_P \equiv G_N^{-1/2}$.

interactions. Already by 1920, Ernest Rutherford had speculated that atomic nuclei might consist of protons and what he termed "neutrons," which however consisted of a proton and electron that were somehow much more tightly bound together than were nuclei and atomic electrons. This was economical in terms of particle content,[5] but soon ran into difficulties as the ideas of quantum mechanics were developed. For example, the observed nuclear sizes and binding energies conflicted with the Heisenberg uncertainty relation, and the model required that the ^{14}N nucleus should have half-integer spin, while molecular spectroscopy indicated that it has spin-1.

These difficulties evaporated with James Chadwick's discovery of the neutron in 1932, which has spin-1/2 rather than the integer spin of Rutherford's bound state. The modern view of the nucleus as consisting of tightly bound protons and neutrons was quickly accepted. Heisenberg and others postulated what is now called *isospin* or $SU(2)$ symmetry. This was the first appearance of an *internal symmetry*, a notion that was central to later developments in particle physics. It implies that the strong interactions of the proton and neutron (known collectively as *nucleons*) are closely related and that they would have the same mass in the absence of electroweak interactions, which do not respect the symmetry.[6]

In 1934, Hideki Yukawa wrote his first paper on what is now called the *Yukawa theory* of the strong interaction.

---

[5]This is an early instance of the reluctance of physicists to invent new particles.

[6]It is now understood that the quark mass differences also contribute to isospin breaking.

The idea was that the strong force is mediated by the exchange of a massive electrically charged *meson*, leading to a pn potential $V(r) \propto g_\pi^2 e^{-m_\pi r}/r$, where g_π is the p-n-meson coupling and m_π is the meson mass. The force can therefore be large for small distance, but falls off rapidly with r, with range $R = 1/m_\pi$. From the observed $R \sim 1$ fm $= 10^{-13}$ cm, Yukawa estimated $m_\pi \sim 200\, m_e$, where $m_e \sim 0.511$ MeV is the electron mass. A few years later, a charged particle with roughly that mass was observed in cosmic rays. However, it gradually became clear that it was not Yukawa's particle, but is actually the *muon* ($\mu^\pm$), a heavier carbon copy of the $e^\pm$. Yukawa's particles, now known as *pions* ($\pi^\pm$), were finally observed in cosmic rays in 1947, and a third electrically neutral π^0 in 1950. They have spin-0 but couple to nucleons as pseudoscalars rather than scalars. The charged pions have mass $m_{\pi^\pm} \sim 270\, m_e$ but are much lighter than the proton mass, $m_{\pi^\pm} \sim 0.15\, m_p$. They decay to the somewhat lighter $\mu^\pm$ and a neutrino by weak interactions, and the π^0 to 2γ electromagnetically. Including heavier mesons discovered later, the appropriately modified Yukawa potential gives an approximate description of the long-range part of the nuclear interaction. Unfortunately, attempts to turn the Yukawa interaction into a full-fledged field theory in the 1950s were not very successful. Unlike QED, which can be expanded perturbatively in the fine structure constant $\alpha = e^2/4\pi \sim 1/137$, the Yukawa interaction is very strong. The analog of α is $g_\pi^2/4\pi = \mathcal{O}(10)$, so that perturbative calculations are not very meaningful.

In addition to the pions, heavier mesons, now known as *kaons* ($K^\pm$, K^0, $\bar{K}^0$), with masses intermediate between the pions and nucleons, were observed to be produced in cosmic ray interactions. Heavier versions of the nucleons (the *hyperons*) were discovered somewhat later. These had the surprising property that they could be produced rapidly by strong interactions, but decayed very slowly, even when there were final states such as $K^+ \to \pi^+\pi^0$ that could presumably be reached by strong processes. The resolution, due to Murray Gell-Mann, Kazuhiko Nishijima, and others, was that these new particles carry a new quantum number, dubbed *strangeness* (S), which is conserved by the strong interactions. Ordinary pions and nucleons have $S = 0$, but the strange particles can be produced strongly by the *associated production* of an $S = 1$ particle, such as K^+ or K^0, and an antiparticle with $S = -1$, such as K^- or $\bar{K}^0$, e.g., $\pi^0 p \to K^+ \bar{K}^0 n$. However, the $K^+ \to \pi^+\pi^0$ decay would violate strangeness[7] and can proceed only by the much more feeble weak interaction, which does not conserve S.

The muons and strange particles did not seem to play any essential role in nature.[8] These were the first observations of particles from a heavier *family*, the role of which is still not fully understood.

Over a hundred additional strongly interacting particles (known as *hadrons*) were discovered at particle accelerators during the 1950s and 1960s, causing Enrico Fermi to remark "If I could remember the names of all these particles, I'd be a botanist." Much was known empirically about their properties and interactions. They had

[7] Strange particle decays also violate isospin.

[8] The muon discovery led Isidor Rabi to make his famous remark, "Who ordered that?"

spins ranging from 0 to 2; fell into isospin multiplets of $2I + 1$ particles with similar masses (but different electric charges), with I from 0 to 3/2; and had integer strangeness 0, ± 1, ± 2, ± 3. They also had well-defined properties under *space reflection (parity)*, in which a physical system is replaced by its mirror image, and under *charge conjugation*, where particles and antiparticles are interchanged. Some hadrons decayed rapidly,[9] respecting the strong interaction symmetries, such as $\rho^0 \to \pi^+\pi^-$, where ρ^0 is a spin-1 particle with mass 775 MeV (80% of m_p). Others could only decay much more slowly by weak or electromagnetic interactions.

Gell-Mann and Yuval Ne'eman independently introduced the *eightfold way* in 1961, an extension of isospin symmetry known as $SU(3)$. Particles of different strangeness and isospin were associated in $SU(3)$ multiplets of dimension 1, 8, 10, and 27. Although $SU(3)$ is broken by some 25% in the masses, the theory greatly simplified the hadron spectrum, led to many successful relations involving the decays and mass differences, and was dramatically confirmed by the successful prediction of a previously unknown particle and its mass, the Ω^-, with spin-3/2 and strangeness -3.

Several years later, Gell-Mann and George Zweig independently introduced the quark model,[10] according to which the known hadrons are bound states of point-like fractionally charged spin-1/2 particles, with the *baryons* (nucleons and hyperons) consisting of three quarks, and

---

[9]With a typical lifetime $\tau \sim 10^{-23}$ s, as inferred from the width $\Gamma = 1/\tau$.

[10]The name was taken from "Three quarks for Muster Mark!" in James Joyce's *Finnegan's Wake*.

the pions and other mesons made from a quark-antiquark pair. Only three types of quarks[11] were needed to account for the hundreds of hadrons, the up (u), down (d), and strange (s), with electric charges of $2/3$, $-1/3$, and $-1/3$, respectively, in units of the positron charge e. The quark model justified $SU(3)$ (the u, d, and s are related in a three-dimensional multiplet), and went beyond it in classifying the hadron states and relating and understanding their intrinsic properties and decays. Nevertheless, all attempts to observe the fractionally charged quarks at accelerators or in cosmic ray experiments were unsuccessful.

In addition to the quark model and $SU(3)$, there was considerable theoretical work in the 1960s attempting to understand the actual dynamics of the strong interactions, motivated by the large body of experimental observations of the hadronic interactions and decays. For example, there were various phenomenological models that described aspects of the strong interaction in certain limits, such as the Yukawa theory for nuclear binding. Other models focused on other aspects, e.g., Regge theory for high-energy scattering, or the more general S-matrix theory, which emphasized general principles of scattering amplitudes[12] such as unitarity (the conservation of probability) and their singularity structure (e.g., poles and branch points in their dependence on kinematic parameters). Other ideas (e.g., current algebra) emphasized the formal structure of $SU(3)$,

---

[11]The number has now proliferated to 6 or 18, depending on how the counting is done.

[12]One class of models based on S-matrix theory, the dual-resonance models, were eventually reformulated as string theories for the strong interactions, and later reinterpreted as string theories for all interactions including gravity.

its extensions to higher symmetries, and the implications for weak and electromagnetic processes. Another thrust was axiomatic field theory (Streater and Wightman 2000), which put field theory on a formal mathematical basis and led to rigorous proofs of the CPT and spin-statistics theorems. These approaches all had limited successes, but none led to a fundamental quantitative understanding of the strong interactions.

The weak interaction is responsible for β decay, i.e., the decay $n \rightarrow pe^-\bar{\nu}_e$, where the neutron can be free or bound in a nucleus, or $p \rightarrow ne^+\nu_e$ if this is energetically allowed due to nuclear binding energies. ν_e and $\bar{\nu}_e$ are respectively the electron-type neutrino and antineutrino, which differ in a conserved (or almost conserved) internal charge known as *lepton number* (L). The neutrinos are almost massless and have no electric charge, so they were invisible to the early experiments. However, they carry off energy, which therefore appeared to be nonconserved in β decay. Pauli suggested in his famous "Dear radioactive ladies and gentlemen" letter of 1930 (Pais 1986, p. 315) that energy conservation could be saved if an unobserved light particle was emitted. In 1934, Fermi published his theory of the β decay interaction, which incorporated Pauli's particle (which Fermi named the neutrino). The *Fermi interaction* was modeled loosely after QED, but instead involved the interaction of four fermions at zero range. Unlike QED and nuclear forces, it allowed the transition of one type of particle into another (e.g., $n \rightarrow p$). Furthermore, it viewed the $e^-\bar{\nu}_e$ pair as being created at the time of the transition, as opposed to having somehow been hiding in the nucleus beforehand.

The Fermi theory has been repeatedly modified to take into account muons, strange particles and strangeness violation, and heavier particles, and has been rewritten in terms of quarks. Even today, it is an excellent first approximation to large numbers of weak interaction decays and other low-energy processes (Commins and Bucksbaum 1983). The existence of the neutrino was directly verified in 1956 when Fred Reines and Clyde Cowan observed the inverse reaction $\bar{\nu}_e \, p \rightarrow e^+ n$ in a target located near a reactor. A second neutrino (ν_μ), associated with the muon in weak transitions such as $\nu_\mu \rightarrow \mu^-$, was discovered at Brookhaven National Laborarory in 1962.

A major conceptual change occurred in the mid-1950s, when T. D. Lee and C. N. Yang realized that space-reflection invariance (or parity, P) and charge-conjugation invariance (C), both of which are respected by QED and the strong interactions, had never been tested for weak interactions.[13] Parity violation was subsequently observed by C.-S. Wu in polarized ^{60}C decay, and C noninvariance soon thereafter. Richard Feynman, Gell-Mann, and others showed that the observations could be accounted for by replacing the vector currents of the Fermi theory by *vector minus axial vector* ($V - A$) currents. In the modified theory C and P are violated maximally, but the product CP is an exact symmetry.

To understand this better, the $V - A$ current only acts on *left-chiral* quarks and leptons and on *right-chiral* antiquarks and antileptons. The precise definition of

---

[13]Parity was apparently taken for granted by most physicists, with the exception of Dirac, who "did not believe in it" (Pais 1986, p. 25).

left- or right-chirality is rather technical, but for our purposes it suffices that chirality coincides with *helicity* when the particle is relativistic, where left- or right-helicity spin-1/2 particles are defined as those with their spin antiparallel or parallel to their momentum, respectively. In the $V - A$ theory, the weak interactions act equally on left-chiral electrons (e_L^-) and right-chiral positrons (e_R^+), while the right-chiral e_R^- and left-chiral e_L^+ are blind to the weak interactions. The physics of e_L^- and e_R^-, which are mapped onto each other by space reflection, are therefore different. Similarly, the physics of the charge conjugates e_L^- and e_L^+ differ, but the CP conjugates e_L^- and e_R^+ should be identical.

CP invariance holds to an excellent approximation, but to the amazement of most physicists a tiny violation of CP invariance was observed in 1964 in a rare kaon decay. The underlying strength is about 10^{-3} compared to ordinary weak amplitudes, and it was unclear whether the CP violation was due to an entirely new interaction. Andrei Sakharov subsequently pointed out that CP violation is one of the necessary ingredients in any dynamical origin of the excess of matter over antimatter in the Universe (the *baryon asymmetry*) (Sakharov 1967).

The Fermi theory could be modified to accommodate C and P violation, but there remained a seemingly insurmountable difficulty. Though it worked well for low-energy processes, it was known that scattering cross sections would grow with energy, becoming so large as to violate unitarity at a center of mass (CM) energy of around 1 TeV $\sim 1000\, m_p$. This was a symptom of the nonrenormalizability of the theory. The higher orders in perturbation theory that would be needed to cure the unitarity problem did not make sense mathematically.

Thus, at the end of the Era of Exploration there was a satisfactory theory only of electromagnetism. Much was known about the properties of the strong and weak interactions, and there were many models that did (and still do) describe aspects reasonably well in appropriate limits. However, there were no consistent fundamental theories and little apparent prospect of any being developed in the foreseeable future. Finally, it was known that classical general relativity was very successful for gravity, but there was no quantum-mechanical version.

The situation was to change dramatically in the next decade.

2.4 The Standard Model Era

The standard model of the strong, weak, and electromagnetic interactions was developed and established in the late 1960s and the 1970s, though it incorporated many elements proposed earlier.

A key ingredient was *Yang-Mills* theory, first proposed in 1954, in which interactions are mediated by the exchange of (apparently massless) spin-1 *gauge particles*. Yang-Mills theories generalize QED. However, unlike the photon, which is electrically neutral, the gauge bosons themselves carry Yang-Mills charges and therefore have elementary self-interactions.

The electroweak part of the standard model, proposed in 1967 by Steven Weinberg (Weinberg 1967) and independently by Abdus Salam (Salam 1968), combines the Yang-Mills interactions with QED, the Fermi theory, and its *intermediate vector boson* extension (in which the weak interactions are mediated by the exchange of massive charged spin-1 particles called the $W^{\pm}$ *bosons*). The gauge interactions are associated with the group $SU(2) \times U(1)$. The weak interaction gauge bosons and *chiral fermions* (i.e., their interactions are parity-violating) are given mass by the *Higgs mechanism* through their interactions with a background spin-0 field, which spontaneously breaks the gauge symmetry but was postulated to preserve renormalizability.

The renormalizability of the model was proved a few years later. Weinberg's original 1967 model was for leptons only. A natural extension to quarks utilizing just the three types that were then known (the u, d, and s) led to predictions for the neutral kaons that were inconsistent with experiment. This could be remedied by postulating a fourth (*charm* or c) quark (the *GIM mechanism*), and in fact the c quark was discovered with roughly the predicted mass a few years later. The c quark also put the quarks and leptons on a similar footing, with four each.

In addition to the *weak charged current* (WCC) interactions mediated by the $W^{\pm}$ (responsible for weak decays), the $SU(2) \times U(1)$ model predicted a new *weak neutral current* (WNC) interaction mediated by a new massive neutral spin-1 particle, the Z *boson*. These neutral current interactions were subsequently observed. In the following decades, the W and Z particles were produced directly with the predicted masses, and the WNC interactions and the properties of the W and Z were measured to high precision, in excellent agreeement with the theoretical predictions (see, e.g., Langacker 2010). The Higgs mechanism was confirmed by the observation of the Higgs boson (i.e., the quantum excitation of the Higgs field) at the Large Hadron Collider (LHC) in Switzerland in 2012, completing the verification of the electroweak theory.

The $SU(2) \times U(1)$ model with only two fermion families, i.e., $(u, d; v_e, e^-)$ and $(c, s; v_\mu, \mu^-)$, does not have any mechanism for CP noninvariance. However, the extension to three families allows the possibility of observable complex phases in weak transitions, implying CP violation. The third-family particles were eventually observed, with the mass of the heaviest quark (the *top*, t) successfully predicted prior to its direct observation through its effect on higher-order weak processes. Many observations of CP violation involving K mesons and their heavy quark analogs are consistent with this origin, although it is not sufficient to account for the baryon asymmetry. The original electroweak model (like the Fermi theory) assumed that the neutrinos are exactly massless. However, by 1998 it was established that the neutrinos have tiny masses, solving the *Solar neutrino problem* (the observed deficit of v_e's produced by nuclear processes in the core of the Sun) by conversion of most of the v_e's into other types. The properties of the neutrino masses and mixings (e.g., whether the masses violate lepton number conservation) continue to be actively studied.

The strong interaction part of the standard model, quantum chromodynamics (QCD), was developed in the early 1970s, strongly motivated by the *deep inelastic electron scattering* experiments at the Stanford Linear Accelerator

Center (SLAC). These indicated that the proton and neutron, which in some sense are "big and fuzzy" with radii ~1 fm, consist of point-like constituents that interact relatively weakly at high energies. The kinematic distributions indicated that these constituents have spin-1/2, tying in nicely with the quark model that had been developed earlier to describe hadron spectroscopy. The quark model, extended by a new three-valued quantum number known as *color*, was combined with $SU(3)$ Yang-Mills theory to give QCD. QCD also led to a simple understanding of the observed strong interaction *flavor*[14] symmetries such as isospin, the eightfold way, and their extensions. QCD was subsequently verified by experiments such as e^+e^- annihilation into hadrons and studies of heavy quark spectroscopy, which observed the effects of the *gluons* (the spin-1 gauge particles), the color quantum number, hadronic jets, and the running of the strong gauge coupling. *Lattice* calculations, in which space and time are approximated by a discrete finite lattice of points, eventually allowed even more detailed tests of the low-energy consequences of QCD and gave insight into why isolated quarks and gluons are not observed.

The standard model combines QCD, the electroweak theory (generalized to include neutrino mass), and classical general relativity. It is a mathematically consistent description of nature. Experiments sensitive to physics on scales

---

[14]Flavor refers to the type of quark, e.g., u, d, or s, each of which comes in three colors, or to the type of lepton. The eightfold way flavor symmetry is based on $SU(3)$. It is mathematically similar to but physically distinct from the QCD color symmetry.

smaller than 1/1000th the size of the atomic nucleus have verified most aspects, often to high precision.

2.5 Beyond the Standard Model

The standard model has been extremely successful, far more so than could have been anticipated in 1967. Nevertheless, it is almost certainly not the final story: it is very complicated, has many free parameters and fine tunings, and leaves a number of questions unanswered, as described in chapter 1.

There are many theoretical ideas for embedding the standard model in a more complete and fundamental theory. One possibility is *unification*, i.e., that two or more seemingly different interactions are really different aspects of a more fundamental and simpler underlying theory. The most familiar example is electromagnetism. The electric and magnetic forces appear to be very different animals, but it was shown in the nineteenth century by James Clerk Maxwell and others that they are really different manifestations of electromagnetism. Similarly, the electroweak part of the standard model at least partially unifies the weak and electromagnetic interactions. Much more ambitious but still unestablished is grand unification, which would unify the strong and electroweak interactions. Even more so are superstring theories, which unify gravity as well.

Grand unification and superstring theories may involve relatively weak couplings all the way up to the Planck scale, $M_P = G_N^{-1/2} \sim 10^{19}$ GeV, the scale at which quantum gravity effects become important. Possible experimental implications of such theories include supersymmetry (a relation between fermions and bosons) at the TeV (10^3 GeV) scale or higher; other low-scale *remnants*, such as new particles or interactions or more complicated versions of the Higgs mechanism; proton decay; the unification of gauge couplings when extrapolated to high energies; or possible large and/or warped extra space dimensions.

There are other possible extensions of the standard model, such as those with *strong coupling* effects at or near the TeV scale. For example, the Higgs boson could be composite and only appears to be elementary when probed at lower energies. Such theories may again lead to new particles and interactions.

Despite all of these theoretical ideas, no direct evidence for supersymmetry, strong coupling, or new particles was observed during the first LHC run, which ended in 2012. Perhaps such ideas are still valid but the mass scales are somewhat higher, or perhaps nature is more subtle,[15] with new physics totally different than we have imagined or even with no new physics at all up to the Planck scale. In any case, one can hope that a variety of experimental and observational probes will combine with theoretical advances to shed light on these issues and possibly even allow an extension of our understanding to the Planck scale. These include high-energy collider experiments at the LHC and at possible future e^+e^- and pp colliders, experiments probing neutrino masses and properties, flavor physics (e.g., involving heavy quarks, rare decays,

[15] "You come to nature with all your theories, and she knocks them all flat" (Pierre-Auguste Renoir).

CP violation, or electric dipole moments), as well as astrophysics (e.g., core-collapse supernova explosions) and cosmology (such as the nature of the dark matter and dark energy, and the origin of the baryon asymmetry).

We now turn to a more detailed investigation of the standard model, its successes and shortcomings, and possibilities for the future.

3

PARTICLES, INTERACTIONS, AND COSMOLOGY

3.1 The Fundamental Particles

According to the quantum-mechanical rotation group, each type of elementary particle must have an intrinsic angular momentum (spin) $S = 0, \frac{1}{2}, 1, \frac{3}{2}, \cdots$ in units of $\hbar$, and can have $2S + 1$ orientations with respect to a given axis. The spin-statistics theorem of quantum field theory implies that particles with integer S must be *bosons*, i.e., the wave function of two identical bosons is symmetric under particle interchange. Particles with half-integer S are *fermions*, with antisymmetric wave functions, implying the Pauli exclusion principle. The known elementary particles are listed in table 3.1.

At our present level of understanding, the fundamental fermions are the quarks, which feel the strong interaction, and the leptons,[1] which do not. They appear to be

---

[1] The term *lepton* derives from a Greek word meaning "small." The original definition of lepton as a fermion with small mass was appropriate when only the e, μ, and neutrinos were known, but is a misnomer for the τ. Nevertheless, it is convenient to keep the term and change the definition.

Table 3.1. The known elementary particles.

spin-0	H			Higgs boson
spin-1/2	$\begin{pmatrix} u_r \\ d_r \end{pmatrix}_L \begin{pmatrix} u_g \\ d_g \end{pmatrix}_L \begin{pmatrix} u_b \\ d_b \end{pmatrix}_L$ $u_{rR} \quad u_{gR} \quad u_{bR}$ $d_{rR} \quad d_{gR} \quad d_{bR}$ quarks		$\begin{pmatrix} \nu_e \\ e^- \end{pmatrix}_L$ $\nu_{eR}\,(?)$ e^-_R leptons	1st family
	$(c, s; \nu_\mu, \mu^-)$		$(t, b; \nu_\tau, \tau^-)$	2nd & 3rd families
	$(\bar u, \bar d; \bar\nu_e, e^+)$	$(\bar c, \bar s; \bar\nu_\mu, \mu^+)$	$(\bar t, \bar b; \bar\nu_\tau, \tau^+)$	antiparticles
spin-1	γ photon	$W^\pm, Z$ weak	G 8 gluons	gauge bosons
spin-2	$g\ (?)$			graviton

Note: L and R refer to left- and right-chiral, respectively; while r, g, and b are the QCD colors. (The right-chiral neutrinos and the graviton are not definitely established.) The strong interactions can transform one color into another. The traditional weak interactions (like β decay) can transform one L quark or lepton into its partner, while leaving the R ones alone. This is reversed for the antiparticles. The heavier fermion families are similar to the first. The spin-0, 1, and 2 bosons are either their own antiparticles, such as γ, or occur in pairs, such as W^+ and W^-.

point-like, i.e., not composites, of smaller particles[2] and are treated as such in the standard model.

There are actually six types of quarks, known as flavors. They have identical strong interactions, but differ in their masses and their electroweak properties. Each of the six have two chiralities, L and R, associated with their spin orientations (page 20), and each comes in three QCD colors, red (r), green (g), and blue (b). These are simply convenient labels, chosen in analogy with but unrelated to optical colors. The colors are actually charges that act as the sources of the strong interaction fields. They are analogous to electric charges in electrodynamics, but are more complicated in that there are three different types.

The six flavors are referred to as up (u), down (d), charm (c), strange (s), top (t), and bottom (b), where the names are partially whimsical but are also motivated by mathematical notation. There are similarly six elementary leptons, the charged electron (e^-), muon (μ^-), and tau (τ^-), and three associated neutrinos ν_e, ν_μ, and ν_τ. Each of these has an antiparticle, denoted generically by $\bar{u}$, $\bar{d}$, $\bar{\nu}_e$, and e^+ for the first family. However, when referring more specifically to a particle of definite chirality, it is convenient to refer to the CP conjugate, which has the opposite chirality but similar interactions, using a slightly different notation. Thus, u_R^c is the right-chiral anti-up quark, the CP conjugate of u_L, and similarly for $u_L^c \leftrightarrow u_R$, $d_{R,L}^c \leftrightarrow d_{L,R}$, $e_{R,L}^+ \leftrightarrow e_{L,R}^-$, and $\nu_R^c \leftrightarrow \nu_L$. (It is not certain whether $\nu_L^c \leftrightarrow \nu_R$ exists.)

---

[2]Compositeness is further discussed in chapter 6.

Table 3.2. The mass, spin, and charge of the known fundamental particles and a few of the bound-state hadrons.

Particle	Mass (GeV)	Spin ($\hbar$)	Electric charge (e)
H (Higgs)	125	0	0
u (up)	0.002	$\frac{1}{2}$	$\frac{2}{3}$
d (down)	0.005	$\frac{1}{2}$	$-\frac{1}{3}$
c (charm)	1.3	$\frac{1}{2}$	$\frac{2}{3}$
s (strange)	0.095	$\frac{1}{2}$	$-\frac{1}{3}$
t (top)	173	$\frac{1}{2}$	$\frac{2}{3}$
b (bottom)	4.7	$\frac{1}{2}$	$-\frac{1}{3}$
e (electron)	0.00051	$\frac{1}{2}$	-1
μ (muon)	0.11	$\frac{1}{2}$	-1
τ (tau)	1.8	$\frac{1}{2}$	-1
$\nu_{e,\mu,\tau}$ (neutrinos)	$\lesssim 10^{-10}$	$\frac{1}{2}$	0
γ (photon)	0	1	0
$W^{\pm}$	80.4	1	± 1
Z	91.2	1	0
G (gluons)	0	1	0
g (graviton)	0	2	0
$\pi^{\pm}, \pi^0$ (pions)	0.14	0	$\pm 1, 0$
$K^{\pm}, K^0, \bar{K}^0$ (kaons)	0.49	0	$\pm 1, 0, 0$
p, n (nucleons)	0.94	$\frac{1}{2}$	$1, 0$
$\rho^{\pm}, \rho^0$	0.77	1	$\pm 1, 0$
$\Delta^{++}, \Delta^{+}, \Delta^{0}, \Delta^{-}$	1.23	$\frac{3}{2}$	$2, 1, 0, -1$

Note: The quark masses listed are due to the Higgs mechanism. They also receive a common mass ~ 0.3 GeV from the strong interactions.

The masses and electric charges of the elementary fermions are listed in table 3.2. It is seen that the quarks and charged leptons have an enormous range of masses, from $\sim 5.4 \times 10^{-4}$ of the proton mass, m_p, up to $\sim 186\, m_p$. The neutrinos are very much lighter.

The first family of quarks and leptons consists of the u, d, ν_e, and e^-. These are all that are needed for ordinary matter. For example, protons and neutrons are bound states $p = uud$ and $n = udd$, while $\pi^+ = u\bar{d}$, $\pi^- = d\bar{u}$, and $\pi^0 = (u\bar{u} - d\bar{d})/\sqrt{2}$.

The second and third families, $(c, s; \nu_\mu, \mu^-)$ and $(t, b; \nu_\tau, \tau^-)$, are identical to the first, except for their larger masses and the decays that are therefore kinematically allowed. The s quark carries the strangeness quantum number, $\mathcal{S} = -1$, so that $K^+ = u\bar{s}$, $K^0 = d\bar{s}$, $K^- = s\bar{u}$, and $\bar{K}^0 = s\bar{d}$. The c, b, and t are heavier still. The heavy families greatly complicate the standard model. It is still not understood why they exist or why the fundamental fermions have such an enormous range of masses.

The strongly interacting particles (hadrons) are all neutral under color. The baryons have half-integer spin and are made of three quarks, with a wave function that is antisymmetric in the three colors. The mesons consist of a quark-antiquark pair, with the colors in the combination color-anticolor.[3] The large number of hadrons is due to the possible total quark spin, radial and orbital excitations, and various flavor combinations. Strong interaction *resonances* are particles with a large width (uncertainty) in the total energy and a very short lifetime. They can decay by rearranging the spin and spatial quantum numbers without any quark changing its flavor, or by the creation of a $q\bar{q}$

---

[3]There is recent evidence for more complicated color-neutral hadrons, such as $qqqq\bar{q}$ baryons and $qq\bar{q}\bar{q}$ mesons.

pair. For example, the Δ^+ is a *uud* state that resembles the proton except that the quark spins add up to $3/2$. It shows up as a very broad *Breit-Wigner* resonance (enhancement in the cross section) for reactions such as $\pi^+ n \to \Delta^+ \to \pi^0 p$, with a width of $\Gamma \sim 120$ MeV and a corresponding lifetime $1/\Gamma \sim 6 \times 10^{-24}$ s. Hadrons that can decay only by changing the flavor of a quark or that are otherwise stable under the strong interactions can decay much more slowly by the weak or electromagnetic interactions. The proton is an exception: it appears to be absolutely stable,[4] or at least to have a lifetime in excess of 10^{31} yr.

The quarks and leptons are sometimes referred to as matter particles. They obey the Pauli exclusion principle and have more or less fixed numbers. A hydrogen atom, for example, consists of an electron and proton (i.e., three quarks) in first approximation. There are also a number of fundamental bosons. These are just as real as matter particles, but do not obey the exclusion principle. For example, the photon (γ), the massless spin-1 quantum of the electromagnetic field, can be freely emitted or absorbed by an electrically charged particle. At its most basic level, the electromagnetic force is due to the exchange of photons.[5] This exchange process is conveniently illustrated by the *Feynman diagram* in the left-hand side of figure 3.1. The vertices and lines in such diagrams encode the quantum-mechanical amplitude for the exchange, as will be described in chapter 4.

---

[4]Possible proton decay is one of the critical issues in particle physics. Most ideas about unification predict that the proton does decay with a very long lifetime, implying the ultimate instability of ordinary matter.

[5]These may be *virtual*, i.e., temporarily violating the relation $E^2 = \vec{p}^{\,2} + m^2$ between energy, momentum, and mass, as allowed by quantum uncertainty.

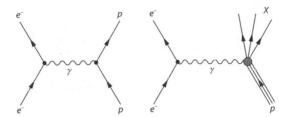

Figure 3.1. Left: The exchange of a photon (γ) between an electron and a proton. The diagram represents both emission from the electron followed by absorption by the proton and the reverse. A classical electromagnetic field consists of a superposition of multiphoton states. Right: An inelastic scattering process, with the proton pictured as three quarks.

There are other spin-1 gauge bosons or force carriers analogous to the photon: the $W^{\pm}$ and Z mediate the weak interactions, while eight colored gluons that couple to the QCD color charges are associated with the strong. Similarly, the newly discovered spin-0 Higgs boson is associated with the generation of mass for the $W^{\pm}$, Z, and leptons, and part of the mass of the quarks, and can in principle mediate a force between them. Finally, gravity is presumably ultimately due to the exchange of a spin-2 *graviton*, though no experiment has had the sensitivity to observe it.

3.2 The Interactions

The known interactions are illustrated in figure 3.2.

The strong interactions are responsible for nuclear binding and for energy generation in stars, reactors, and

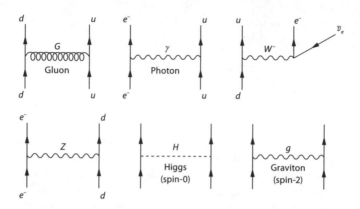

Figure 3.2. The known interactions. The gluons couple to all quark flavors, changing color but not flavor at the vertices. The photon couples diagonally (i.e., without changing the fermion type) to each type of charged particle. The $W^\pm$ vertices are off-diagonal, i.e., they change a quark or lepton into another differing in electric charge, but do not change the quark color. The Z, Higgs, and graviton couple diagonally to each known fermion, with the exception (for the Z) of the right-chiral neutrinos. Spin-0 particles, fermions, electroweak bosons and gravitons, and gluons are represented by dashed, solid, wavy, and curly lines, respectively.

(unfortunately) weapons. They are fundamentally due to QCD, i.e., to the exchange of eight (massless) gluons between quarks. They couple identically to each quark flavor, which is never changed at a quark-gluon vertex. The gluons themselves carry color (unlike the photon, which is electrically neutral), so the quark color can be changed at the vertex and there are additional interactions amongst the gluons. No isolated quarks or gluons have ever been observed, presumably because the forces between

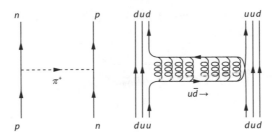

Figure 3.3. Left: The (Yukawa) exchange of a pion between nucleons. Right: Illustration of how the Yukawa interaction emerges from QCD.

colored particles become so great at large distances that they can never be separated (*confinement*). Nevertheless, the evidence for the existence of quarks, gluons, and color is compelling, as will be described in chapter 4.

Since the directly observed strongly interacting particles (nucleons, pions, and other hadrons) are neutral under color, the strong interactions between them are analogous to the dipole forces of atomic and molecular physics. The Yukawa interaction due to pion exchange, mentioned in chapter 2, gives a rough description of the long-range part of the nuclear interaction. The interaction is illustrated in figure 3.3, and some characteristics in table 3.3. It is strong but of very short range.

The electromagnetic interaction acts on all electrically charged particles, including the $e^{\pm}$, $\mu^{\pm}$, $\tau^{\pm}$, p, $\bar{p}$, $\pi^{\pm}$, and quarks. It provides the binding for atoms, molecules, crystals, and other forms of ordinary matter; is responsible for chemical energy, chemical reactions, and modern electronics; and leads to electromagnetic radiation. It is due to

Table 3.3. Characteristics of the known interactions of leptons and hadrons.

	Potential	Strength	Range
Strong	$g_\pi^2 \dfrac{e^{-m_\pi r}}{r}$	$\dfrac{g_\pi^2}{4\pi} \sim 14$	$\dfrac{1}{m_\pi} \sim 10^{-13}$ cm
Electromagnetic	$\dfrac{e^2}{r}$	$\alpha = \dfrac{e^2}{4\pi} \sim \dfrac{1}{137}$	∞
Weak	$g^2 \dfrac{e^{-M_V r}}{r}$	$\dfrac{g^2 E^2}{M_V^2} \sim 10^{-11}$	$\dfrac{1}{M_V} \sim 10^{-16}$ cm
Higgs	$\dfrac{m_1 m_2}{M_V^2} \dfrac{e^{-M_H r}}{r}$	$\dfrac{m_1 m_2}{M_V^2} \sim 10^{-10}$	$\dfrac{1}{M_H} \sim 10^{-16}$ cm
Gravity	$G_N \dfrac{m_1 m_2}{r}$	$G_N m_1 m_2 \sim 10^{-44}$	∞

Note: (The more fundamental QCD interaction between quarks will be characterized in chapter 4.) The amplitudes for emission or absorption of the exchanged boson are g_π, e, g, m/M_V, and m/M_P, respectively, where m is the fermion mass, M_V can represent M_W or M_Z, and M_P is the Planck mass. E is a typical energy in a weak decay, such as the initial mass or the energy release. The numerical examples are for E and $m_{1,2} \sim 1$ MeV. The potentials are qualitative only, and are meant to illustrate the strengths and ranges.

photon exchange as described by QED, is long range due to the masslessness of the photon, and is much weaker than the strong interaction.

The weak interactions affect all known fermions except right-chiral neutrinos. In particular, they are the only significant interaction for the ordinary neutrinos (left-chiral) and antineutrinos (right-chiral). They are too feeble to lead to any kind of binding for the ordinary particles, but are responsible for weak decays. They are critical in the *nucleosynthesis* of elements heavier than hydrogen in the early Universe, stars, and supernova explosions. They also allow scattering processes involving neutrinos and contribute a tiny parity-violating perturbation to the electromagnetic interactions in atoms and charged-lepton scattering. The weak interactions are the only ones that violate space-reflection (parity) and charge-conjugation invariance, because they couple differently to left- and right-chiral fermions. They also violate CP at a much lower level. The weak interactions are now understood to be mediated by the exchange of very massive spin-1 gauge bosons, and are therefore extremely short range. The weak charged current (WCC) is associated with the charged $W^{\pm}$ with mass ~ 80 GeV, known historically as intermediate vector bosons, while the weak neutral current (WNC) is due to the somewhat heavier neutral Z boson. The WCC acts only on L-chiral fermions and can violate strangeness. The Z couples to both L and R, but with different strengths. The weak interactions actually become stronger at higher energies, especially at energies larger than $M_{W^{\pm},Z}$.

The Higgs mechanism generates an effective mass for gauge bosons and chiral fermions via their interaction with a background Higgs field. The Higgs-Yukawa interaction[6] is due to the exchange of the quantum excitation of that field, the spin-0 Higgs boson. The Higgs is very massive, ~ 125 GeV, so the interaction is extremely short range. Its coupling to an elementary fermion of mass m_f is proportional to m_f/v, where $v \sim 246$ GeV is the weak interaction scale that is set by the background Higgs field. This is extremely small for the lighter quarks and leptons.

---

[6]It is conventional to refer to the exchange of the Higgs boson as the "Yukawa interaction" in analogy with the original Yukawa model for the strong interactions. To avoid confusion, it will be referred to here as the Higgs or Higgs-Yukawa interaction.

The gravitational interaction is the weakest of all. Known gravitational effects can be described by general relativity or its Newtonian approximation, but it is presumably due to the exchange of a massless spin-2 graviton at the microscopic level. The graviton couples to all sources of energy, so the strength is proportional to mass for a nonrelativistic particle. The scale is set by the Planck scale $M_P = G_N^{-1/2} \sim 10^{19}$ GeV, so the gravitational interaction for individual particles is miniscule, even compared to the weak and Higgs-Yukawa. However, unlike the strong, electromagnetic, and weak interactions, gravity is always attractive. Furthermore, unlike the Higgs-Yukawa interaction, it is long range. The gravitational force is therefore additive for macroscopic objects: it is responsible for weight, the binding of the Galaxy and Solar System, and the inward force in the Sun that counteracts the thermal pressure created by nuclear fusion.

It is apparent that the interactions have vastly different properties. Particle physics is complicated! One of the underlying goals in physics is to achieve as simple a picture as possible. One possibility is unification, such as the unification of electric and magnetic forces described in section 2.5. Similarly, the standard electroweak theory at least partially unifies the weak and electromagnetic interactions, which would look very much alike if it were not for the symmetry breaking associated with the Higgs field (or if probed at high enough energies). Grand unified theories, which were introduced soon after the standard model, ambitiously attempted to unify all three of the microscopic gauge interactions (QCD and the electroweak theory), with their relation manifest at a scale a few orders of magnitude below the Planck scale. Such theories were elegant and had some intriguing predictions, but at least the simplest versions have now been excluded by the nonobservation of proton decay. They may reemerge, perhaps with the symmetry broken in new *compact dimensions* of space. Even more challenging is the possibility of bringing gravity into the game, perhaps in superstring theory. These ideas will be elaborated in chapter 6.

3.3 Cosmology

Virtually all aspects of classical and modern physics are relevant to astrophysics and cosmology. In particular, the development of the SM was essential for our understanding of the early Universe. Conversely, astrophysics and cosmology frequently yield valuable information about physics, such as the discovery of helium, tests of general relativity, the first indication that neutrinos have a tiny but nonzero mass, the existence of dark matter and energy, and stringent constraints on possible extensions of the standard model. Here I give a brief overview of our current understanding, with a timeline of likely events summarized in table 3.4. More detailed treatments may be found in, e.g., Kolb and Turner 1990; Weinberg 2008; and Loeb 2010.

Two of the most important observations in cosmology were made in the 1920s by Edwin Hubble. He built upon the work of others to establish that many observed nebulae are really (in modern terms) other galaxies far beyond our own Milky Way. Even more important, he used the

Table 3.4. Important cosmological events.

t	T	Event	(Subsequent) composition
10^{-44} s	10^{19} GeV (10^{32} K)	Planck time	quantum gravity (?)
?	?	inflation (?)	
10^{-38} s	10^{16} GeV (10^{29} K)	radiation domination begins	$q, \bar{q}, \ell, \bar{\ell}, G, EW, \phi$
10^{-34} s	10^{14} GeV (10^{27} K)	baryogenesis	
10^{-10} s	100 GeV (10^{15} K)	electroweak transition	$q, \bar{q}, \ell, \bar{\ell}, G, \gamma$
10^{-6} s	1 GeV (10^{13} K)	quark-hadron transition	$e^{\pm}, \mu^{\pm}, \overset{(-)}{\nu}, \gamma$; hadrons
1 s	1 MeV (10^{10} K)	neutrino freezeout	$e^{\pm}, \overset{(-)}{\nu}, \gamma$; (trace) p, n
4 s	0.5 MeV (6×10^9 K)	e^+e^- annihilation	$\overset{(-)}{\nu}, \gamma$; (trace) e^-, p, n
1-180 s	$1-0.1$ MeV (10^{10}–10^9 K)	BBN	$\overset{(-)}{\nu}, \gamma$; (trace) $e^-, p, {}^4He$
10^5 yr	0.7 eV (8000 K)	matter domination begins	$e^-, p, {}^4He$, DM
4×10^5 yr	0.3 eV (3000 K)	recombination	atoms, DM
10^6 yr	6×10^{-3} eV (70 K)	first stars, galaxies	ordinary matter, DM
9×10^9 yr	3×10^{-4} eV (4 K)	DE dominates; Solar System forms	ordinary matter, DM, DE
14×10^9 yr	2×10^{-4} eV (2.7 K)	dinosaurs, Roman Empire, iPhone	ordinary matter, DM, DE

Note: t and T are respectively the time since the big bang and the temperature. T_{RH} is taken at 10^{16} GeV for illustration, and the baryogenesis temperature corresponds to one particular model (*leptogenesis*). ℓ, EW, ϕ, DM, and DE refer respectively to leptons, electroweak bosons, the Higgs doublet, dark matter, and dark energy. New (beyond the standard model) particles may also have been present. ν, $\bar{\nu}$, and γ were present but subdominant in the latter periods.

observed redshifts to show that the distant galaxies are receding from us, suggesting that the Universe as a whole is expanding. One can work backward to infer that the observed Universe must have started with an extremely (or infinitely) hot and dense big bang some 14 billion years ago, and has been expanding and cooling since then.[7] *Big bang nucleosynthesis* (BBN) and the *cosmic microwave background radiation* (CMB), to be described later, provide compelling confirmation of the big bang hypothesis.

We do not know what happened at the very first instant. Perhaps the Universe emerged from some kind of phase transition, or it may have bounced back from a previous period of collapse. There was possibly a brief period when the temperature was at the Planck scale and quantum gravity dominated, or perhaps there was nothing but vacuum. In any case, the big bang has two major problems that must be surmounted. First, the CMB is observed to be very uniform in all directions of the sky, even though, for angular separations larger than a few degrees, the regions from which it was emitted were not in causal contact while the radiation was still in equilibrium (the *horizon problem*). Second, the timescale for a homogeneous and isotropic universe to either expand to a maximum and then collapse (if it is closed[8]), or to expand into essentially a

[7] The basic idea of the big bang had been proposed earlier by the scientist and Jesuit priest Georges Lemaître. An alternative explanation, the *steady state* theory, postulated that matter is spontaneously generated to replace the receding galaxies, leading to an expanding but eternal and unchanging Universe. This view was discredited by the late 1960s, however, when observation of very distant objects such as quasars, which emitted their radiation at much earlier times, showed that the Universe is changing.

[8] In a closed universe, the kinetic energy of expansion is smaller than the gravitating energy associated with matter, radiation, and the vacuum. It has a

zero temperature emptiness (if it is open) is the Planck
time ($\sim 10^{-43}$ s). The observed age is some 10^{60} larger,
requiring that the Universe is *flat* to an incredibly fine-
tuned precision (the *flatness problem*).

These difficulties can be resolved if the Universe initially
underwent a brief period of exponentially fast expansion
(*inflation*), in which it was smoothed out and flattened,
and in which the densities of any preexisting particles
were diluted to essentially zero. Inflation could have come
about if the energy density was dominated by vacuum
energy, perhaps associated with some cosmological analog
of the Higgs field. The inflation must eventually have
stopped, with the vacuum energy somehow converted to
a thermal bath of elementary particles with a very high
reheating temperature T_{RH}. During the inflationary phase,
quantum fluctuations would have formed the seeds that
later grew into structures such as galaxies. Inflation has not
been conclusively established, but some form or something
similar seems likely.

We will not speculate further on what might have
occurred prior to inflation or on its details, but will simply
assume that the conventional part of the big bang began,
perhaps following an inflationary phase, at a temperature
T_{RH}, with the appropriate flatness and uniformity. T_{RH} is
unknown, but it is usually assumed that it is smaller than
the Planck scale[9] (so that quantum gravity can be ignored)
and well above a few MeV (to account for BBN).

---

curvature similar to the surface of a sphere. In an open universe, the kinetic
energy is larger than the gravitating energy, with curvature analogous to a saddle.
A flat universe is at the border, with equal kinetic and gravitating energy. It is
analogous to an object thrown upward at the escape velocity.

[9]Temperature and energy are related by the Boltzmann constant, 8.6×10^{-5} eV/K.

Assuming, as is likely, that $T_{RH} \gg \mathcal{O}(100 \text{ GeV})$, the electroweak symmetry would have been initially unbroken and the known particles would all have been massless. The Universe would then have consisted of an almost uniform plasma of quarks, leptons, antiquarks, antileptons, photons, gluons, electroweak bosons, Higgs particles, and (possibly) particles associated with new physics, in a period known as the *radiation-dominated* epoch. Radiation refers here to all relativistic particles, both bosons and fermions. The energy density in radiation scales as T^4, where T is the temperature. As the Universe expanded, the particle wavelengths were redshifted so that the temperature decreased, i.e., $T \propto R^{-1}$, where R is the *scale factor* characterizing the expansion.

The details of this epoch depend to some extent on unknown physics, but presumably the baryon asymmetry (excess of matter over antimatter) was somehow generated (*baryogenesis*). The baryon asymmetry and some possible mechanisms for its origin will be described in some detail in chapter 5, but the key point is that the amount of matter relative to photons observed today implies that there must have been a slight excess (one part in 10^9) of quarks with respect to antiquarks early in the radiation period, which must have been generated dynamically after or during the reheating following inflation. As the plasma cooled, there would have been several events that our present understanding of particle physics allows us to speculate on with a reasonable degree of confidence. The *electroweak phase transition* presumably occurred at $T = \mathcal{O}(100 \text{ GeV})$, when it became sufficiently cool for a classical Higgs field to develop, breaking the electroweak symmetry and turning on masses. As T dropped even more, various heavy particles would have annihilated against their antiparticles, dumping their energy and entropy into lighter species. The *quark-hadron transition* took place at a few hundred MeV. At higher temperatures, quarks, antiquarks, and gluons were unconfined free particles. Below that temperature, most of the quarks and antiquarks annihilated each other. Eventually, only the small excess of quarks survived, and these were bound into nucleons. By the time that T fell to $\mathcal{O}(10 \text{ MeV})$, the plasma consisted mainly of electrons, positrons, neutrinos, antineutrinos, and photons. There were also trace amounts of protons, neutrons, and (presumably) other particles that would make up the dark matter. These were still negligible players in the dynamics of the Universe, but would soon come to dominate.

At sufficiently high temperatures, the weak interaction processes

$$n + \nu_e \leftrightarrow p + e^-, \qquad n + e^+ \leftrightarrow p + \bar{\nu}_e \qquad (3.1)$$

kept the ratio of neutrons to protons in thermal equilibrium,

$$\frac{n}{p} = \exp\left(-\frac{E_n - E_p}{T}\right) \sim \exp\left(-\frac{m_n - m_p}{T}\right),$$
$$(3.2)$$

where the mass difference is $m_n - m_p \sim 1.29$ MeV. Equation (3.2) held as long as the weak interaction rate $\Gamma \sim G_F^2 \, T^5$ exceeded the Universe's expansion rate (*Hubble parameter*), $H \sim \sqrt{g_*} \, T^2 / M_P$. Here, $G_F = \sqrt{2} g^2 / 8 M_W^2 \sim 1.2 \times 10^{-5} \text{ GeV}^{-2}$ is the *Fermi constant*, describing the weak interaction strength, M_P is the Planck scale, and g_* ($= 43/4$ for $m_e < T < m_\mu$) is related to the number of relativistic degrees of freedom in the plasma. However, Γ dropped below H at the *neutrino freeze-out temperature*[10] $T_f \sim (\sqrt{g_*}/G_F^2 M_P)^{1/3} = \mathcal{O}(1 \text{ MeV})$. At lower temperatures, the equilibrium could no longer be maintained, so that n/p was fixed at $\exp(-\frac{m_n - m_p}{T_f})$ except for neutron β decay. Most of the neutrons were eventually incorporated into ^4He, leading to the prediction that the ratio by mass of primordial ^4He to H should be $\sim 24\%$, a result that is consistent with the abundances observed in the oldest stars.[11] This big bang nucleosynthesis (BBN) result is quite robust. It depends only on well-known particle, nuclear, and statistical physics, except for the assumption that no unknown sources of radiation were in equilibrium.[12] This nuclear alchemy successfully tests the big bang theory (and constrains extensions of the standard model) when the Universe was only $\sim (1 - 180)$ s old, the earliest period that has been probed directly (Weinberg 1983).

Shortly after the neutrino freezeout, the e^+ annihilated with e^- producing photons, leaving radiation in the form

[10]It is a remarkable coincidence that T_f, which depends on the strengths of the gravitational and weak interactions, is numerically very close to $\Delta m \equiv m_n - m_p$. The latter receives contributions from the quark masses, which are considered part of the strong interactions but actually generated by the Higgs mechanism, and also from electromagnetism. Thus, all of the known interactions apparently conspire to avoid $T_f \ll \Delta m$ (no primordial ^4He) or $T_f \gg \Delta m$ (no primordial H).

[11]Trace amounts of D, ^3He, and ^7Li were also produced. Heavier elements and additional ^4He were later synthesized in stars, in cosmic ray interactions, and in core-collapse supernovae.

[12]The predicted ^4He to H ratio depends weakly on the nucleon to photon ratio, but that can be determined from other observations involving the BBN and the CMB.

of γ's, ν's, and $\bar{\nu}$'s, as well as trace amounts of excess electrons (analogous to the excess baryons) and nucleons, with the BBN occurring in the next several minutes.

After about 100,000 yr, the Universe had cooled enough to be *matter dominated*, i.e., the energy density in nonrelativistic particles, including ordinary and dark matter, exceeded the radiation energy.[13] By around 380,000 yr, the temperature had dropped enough (to $T \sim 0.3$ eV) for the residual electrons and nuclei to form stable atoms (*recombination*). The remaining photons no longer interacted efficiently enough to stay in thermal equilibrium. They have remained in the Universe ever since, essentially undisturbed except for the expansion of the Universe, which redshifted their wavelengths so that they now have an energy distribution of thermal form with temperature $T_0 = 2.7$ K, in the microwave region. This CMB had been predicted in 1948 by Ralph Alpher and Robert Herman, and was discovered accidentally by Arno Penzias and Robert Wilson in 1964. It is considered to be definitive evidence for the big bang. It is remarkably uniform,[14] but after many earlier observations small temperature fluctuations of $\mathcal{O}(10^{-5})$ were seen by instruments on the COBE satellite in 1992. These formed the seeds from which structures such as galaxies and clusters ultimately formed, and may themselves have been created during an earlier period of inflation. The CMB has been studied in

---

[13]The radiation density falls as $T^4 \propto R^{-4}$, while the matter density falls more slowly, as $T^3 \propto R^{-3}$. (There are corrections to the proportionality between T and R^{-1} at the times of particle-antiparticle annihilation.)

[14]With the exception of a dipole anisotropy associated with the relative motion of the Solar System.

great detail since then, including the recent measurements by the high-precision instruments on the WMAP and Planck satellites, with the details allowing determination of cosmological parameters describing the dark energy, dark and baryonic matter, curvature, and Hubble parameter, and setting limits on neutrino masses and on possible new types of radiation.

Density perturbations could start to grow once the Universe was matter dominated. After $\mathcal{O}(10^8$ yr), the first stars and galaxies formed (which in turn produced heavier elements incorporated in later stars and planets). The details are less understood than those of the CMB and BBN (see, e.g., Loeb 2010), and are a principal focus of current research,[15] including that of the James Webb Space Telescope (successor to the Hubble Space Telescope), scheduled to launch in 2018.

It has been known for decades that there is much more matter in the Universe than the ordinary atoms of which we are made. This *dark matter* does not emit or absorb significant amounts of light. Its existence is inferred from its gravitational effects, e.g., on the motions of stars and gas within galaxies and clusters, from gravitational lensing, from the separation of ordinary and dark matter in galaxy collisions, and from anisotropies in the CMB. It will be further discussed in chapter 5.

The Universe has continued to expand until the present time, some 14 billion years after the big bang. One other event during that period is of particular relevance. The

[15]Large space- and ground-based telescopes are effectively time machines because they observe distant astronomical objects as they were long ago when they emitted their light.

energy density associated with matter falls like $T^3 \propto R^{-3}$ as the Universe expands, while a possible *dark energy* associated with space itself does not vary with R. In the late 1990s, observations of very distant Type I supernovae indicated that the expansion rate of the Universe is actually accelerating. This suggests the existence of dark energy, which acts repulsively at large distances and may be a weaker version of the energy that led to inflation. Subsequent observations have confirmed the existence of the dark energy, which began dominating over the matter density somewhat before the formation of the Solar System 4.6 billion years ago.

There is now a remarkable agreement of multiple observations, including the CMB, Type I supernovae, distributions and dynamics of galaxies and clusters, the BBN, and gravitational lenses. These have established the existence of dark matter and dark energy, and confirmed that the Universe is very close to flat, with the present ratio of energy densities in the form of ordinary matter, dark matter, and dark energy, around 5%, 25%, and 70%, respectively.[16] These ingredients are collectively referred to as the *standard cosmological model*.

What of the future? We are lucky that the Universe is close enough to flat that there was adequate time for us to come into existence, and the timescale for any future cosmological calamity is enormous.[17]

---

[16]The energy in radiation, such as the CMB and neutrinos, is now negligible.

[17]The possibility of a catastrophic decay of a metastable vacuum state will be described in chapter 4.

4

THE STANDARD MODEL

4.1 Gauge Invariance and QED

Gauge invariance emerged historically as an apparently accidental feature of classical electrodynamics. It was maintained in the classical and quantum mechanics of a charged particle in an electromagnetic field, and in QED. One can invert the logic and take a generalized version of gauge invariance as the starting postulate for a physical theory. It turns out that such gauge invariance[1] requires the existence of (apparently) massless spin-1 gauge bosons analogous to the photon, and that the form of the interaction of these gauge bosons with other particles or with themselves is determined. Gauge theories are the unique possibility for renormalizable field theories involving spin-1 bosons, and they provided the breakthrough needed to understand the strong and weak interactions.

---

[1] The coordinate transformations of general relativity can also be viewed as a kind of gauge invariance.

Classical Electrodynamics

The Maxwell's equations of classical electrodynamics are

$$\vec{\nabla} \times \vec{E} + \frac{\partial \vec{B}}{\partial t} = 0, \qquad \vec{\nabla} \cdot \vec{B} = 0,$$

$$\vec{\nabla} \times \vec{B} - \frac{\partial \vec{E}}{\partial t} = \vec{J}, \qquad \vec{\nabla} \cdot \vec{E} = \rho, \tag{4.1}$$

where $\vec{J}$ and ρ are respectively the current and charge densities. We use Heaviside-Lorentz units and have taken $c = 1$. The continuity equation[2] that expresses the conservation of electric charge,

$$\vec{\nabla} \cdot \vec{J} + \frac{\partial \rho}{\partial t}, \tag{4.2}$$

follows directly from (4.1).

It is apparent from the first two equations in (4.1) that the electric and magnetic fields can be expressed in terms of the vector and scalar potentials $\vec{A}(\vec{x}, t)$ and $\phi(\vec{x}, t)$, with

$$\vec{B} = \vec{\nabla} \times \vec{A}, \qquad \vec{E} = -\vec{\nabla}\phi - \frac{\partial \vec{A}}{\partial t}. \tag{4.3}$$

Neither the Maxwell's equations nor the Lorentz force equation

$$\vec{F} = -e(\vec{E} + \vec{v} \times \vec{B}) \tag{4.4}$$

---

[2]Historically, the displacement current $\partial \vec{E}/\partial t$ in the $\vec{\nabla} \times \vec{B}$ equation was introduced by Maxwell to ensure consistency with the continuity equation.

for an electron of charge $-e < 0$ moving in an electro-magnetic field depend explicitly on these potentials, so they are not directly observable. Moreover, they are not unique. The *gauge-transformed* potentials

$$\vec{A}' \equiv \vec{A} + \frac{1}{e}\vec{\nabla}\beta, \qquad \phi' \equiv \phi - \frac{1}{e}\frac{\partial\vec{\beta}}{\partial t}, \qquad (4.5)$$

where $\beta(\vec{x}, t)$ is an arbitrary differentiable function of $\vec{x}$ and t, describe the same $\vec{E}$ and $\vec{B}$ and lead to the same physics. At the classical level, this redundancy in descriptions is mainly a curiosity, although the mathematics is often simplified by choosing β appropriately, e.g., so that $\vec{\nabla} \cdot \vec{A} = 0$.

Quantum Mechanics

The vector and scalar potentials play a more direct role in quantum mechanics, which is formulated in terms of potentials rather than forces. Even though the classical Lorentz force in (4.4) has no explicit dependence on $\vec{A}$ and ϕ, the classical Lagrangian

$$L = \frac{1}{2}m\vec{v}^2 - e\,(\vec{v} \cdot \vec{A} - \phi) \qquad (4.6)$$

does, implying that the canonical momentum conjugate to an electron position $\vec{x}$ is $\vec{P} \equiv m\vec{v} - e\vec{A}$ rather than $m\vec{v}$. Thus, the Hamiltonian becomes

$$H = \frac{1}{2}m\vec{v}^2 - e\,\phi = \frac{(\vec{P} + e\vec{A})^2}{2m} - e\,\phi. \qquad (4.7)$$

In passing from classical to quantum mechanics, it is $\vec{P}$ rather than $m\vec{v}$ that is quantized and identified with $-i\vec{\nabla}$ when acting in position space. Thus, the Schrödinger equation for an electron moving in a classical electromagnetic field becomes (with $\hbar = 1$)

$$\left[-\frac{1}{2m}(\vec{\nabla} + ie\vec{A})^2 - e\phi \right] \psi = i\frac{\partial\psi}{\partial t}. \qquad (4.8)$$

This replacement of $-i\vec{\nabla}$ and $i\frac{\partial}{\partial t}$ in the free-particle equation by the *gauge covariant derivatives* $-i\vec{\nabla} + e\vec{A}$ and $i\frac{\partial}{\partial t} + e\phi$, respectively, is known as the *minimal electromagnetic substitution*.

Equation (4.8) is not by itself invariant in form under the gauge transformation (4.5). However, if in addition we replace $\psi(\vec{x}, t)$ by

$$\psi'(\vec{x}, t) \equiv e^{-i\beta(\vec{x},t)}\,\psi(\vec{x}, t), \qquad (4.9)$$

then gauge invariance is restored. That is, the primed quantities satisfy the same equation as the unprimed ones, because the shifts in $\vec{A}$ and ϕ are compensated by the derivatives acting on the phase. A quantum-mechanical gauge transformation therefore involves both a transformation on the electromagentic potentials and a simultaneous change in the phase of the wave function.

To see what good all of this is, let us work backward, starting from the Schrödinger equation (2.1) on page 9. For simplicity, we consider a free electron, $V(\vec{x}, t) = 0$, although the same considerations apply to a nonelectromagnetic potential. Equation (2.1) is invariant under the replacement $\psi \rightarrow \psi' \equiv \exp(-i\beta)\,\psi$, where β is an arbitrary real constant, i.e., ψ and ψ' satisfy the same equation. These transformations are known as the *global* $U(1)$ group. Global means that it is independent of $\vec{x}$ and t, in contrast with the more general *local* (i.e., gauge) transformations in (4.9), which can be carried out independently at different points in space and time. $U(1)$ refers to the *group* of unitary 1×1–dimensional matrices, which is an incredibly complicated name for phase factors.

Equation (2.1) is *not* invariant in form under local transformations because of the derivatives. However, one can *impose* gauge invariance, either from an esthetic motivation,[3] or, more mundanely, as a convenient prescription. This requires the introduction of vector and scalar potentials, which, subject to a few caveats, must have the equations of motion, interactions, and gauge transformations presented earlier. Thus, a local $U(1)$ gauge invariance largely determines the existence and properties of an associated interaction, up to a coupling constant (charge) e, which unfortunately must be taken from experiment.

Quantum Electrodynamics

Similar considerations hold for the relativistic generalization of the Schrödinger equation (the Dirac equation), but we will jump directly to quantum electrodynamics,

---

[3]It is sometimes argued, for example, that physics should not care whether the phases of an electron on the Earth and one on the "dwarf planet" Pluto are the same. This argument is somewhat undermined because the SM does employ global symmetries. However, these could be remnants of more fundamental local symmetries.

the relativistic field theory that allows the creation and annihilation of photons and of electron-positron pairs (see, e.g., Kinoshita 1990).

Let us first introduce some relativistic notation. The energy-momentum (contravariant) four-vector of a particle of mass m is denoted p, with components $p^\mu = (E, \vec{p})$, where $\mu = 0, 1, 2, 3$. The covariant four-vector is $p_\mu \equiv \sum_{\nu=0}^{3} g_{\mu\nu} p^\nu \equiv g_{\mu\nu} p^\nu$, where $g_{\mu\nu}$ is a diagonal 4×4 matrix with elements $(+1, -1, -1, -1)$, so that $p_\mu = (E, -\vec{p})$. The quantity $p^2 \equiv p^\mu p_\mu = E^2 - \vec{p}^{\,2}$ is Lorentz invariant. The particle is said to be real or on-shell if E and $\vec{p}$ satisfy the physical relation $p^2 = m^2$. Otherwise, it is virtual, i.e., the relation is violated, as can happen briefly because of quantum uncertainty. Similarly, the position four-vector x has components $x^\mu = (t, \vec{x})$ or $x_\mu = (t, -\vec{x})$. Spatial derivatives are conveniently denoted as $\partial^\mu \equiv \frac{\partial}{\partial x_\mu} = \left(\frac{\partial}{\partial t}, -\vec{\nabla}\right)$.

We introduce the electromagnetic vector potential four-vector and field strength tensor

$$A^\mu = (\phi, \vec{A}), \qquad F^{\mu\nu} = \partial^\mu A^\nu - \partial^\nu A^\mu. \quad (4.10)$$

From (4.3),

$$F^{\mu\nu} = \begin{pmatrix} 0 & -E_x & -E_y & -E_z \\ E_x & 0 & -B_z & B_y \\ E_y & B_z & 0 & -B_x \\ E_z & -B_y & B_x & 0 \end{pmatrix}. \quad (4.11)$$

In field theory, it is convenient to introduce the *Lagrangian density*

$$\mathcal{L}_A = -\frac{1}{4} F_{\mu\nu} F^{\mu\nu} = \frac{1}{2}\left(\vec{E}^2 - \vec{B}^2\right). \qquad (4.12)$$

The equations of motion (in this case, Maxwell's equations in empty space) can be derived from $\mathcal{L}_A$, just as the Lorentz force law for the electron follows from (4.6). In the following, we will encounter a number of such Lagrangian densities, which are useful for displaying the gauge invariance and other symmetries, and the associated interactions of the theories. In the present case, we see immediately that $\mathcal{L}_A$ is unchanged by the gauge transformation in equation (4.5), which becomes $A'^{\mu} = A^{\mu} - \frac{1}{e}\partial^{\mu}\beta$ in relativistic notation.

All of this is identical to classical electrodynamics, except that we now interpret A^{μ} as an operator that can act on a state to create or annihilate a photon. Schematically,[4]

$$A \sim \int d^3\vec{k}\left[a(\vec{k}) + a^{\dagger}(\vec{k})\right], \qquad (4.13)$$

where $a(\vec{k})$ and $a^{\dagger}(\vec{k})$ are harmonic oscillator-like annihilation and creation operators. $a^{\dagger}(\vec{k})$ acting on the ground state of the theory produces a one-photon state with momentum $\vec{k}$ and energy $\omega = |\vec{k}|$, while $a(\vec{k})$ acting on that state returns us to the ground state. In (4.13), we have suppressed details involving the photon polarization, and in the following we will not display the $\vec{k}$ dependence.

---

[4]This is the expression in the absence of interactions, which is what is relevant for a perturbative calculation.

The electron wave function can similarly be replaced by
an electron field operator $\psi(x)$, which can either annihilate
an electron or create a positron, i.e., the state $\psi|\chi\rangle$ involves
one fewer e^- or one more e^+ than $|\chi\rangle$, while the adjoint
$\psi^\dagger$ annihilates a positron or creates an electron. Thus,

$$\psi \sim \int d^3\vec{p}\,\left[b(\vec{p}) + d^\dagger(\vec{p})\right], \qquad (4.14)$$

where $b^\dagger(\vec{p})$ creates an electron of momentum $\vec{p}$, $d^\dagger(\vec{p})$
creates a positron, and b and d are the corresponding
annihilation operators.

ψ has four components, not displayed in (4.14), associated with the two spin orientations for e^- and e^+. The
Lagrangian density for a noninteracting electron is

$$\mathcal{L}_e = \bar{\psi}(i\gamma^\mu\partial_\mu - m_e)\psi, \qquad (4.15)$$

which leads to the correct Dirac equation of motion. In
(4.15), γ^μ are the 4×4–dimensional Dirac matrices[5]
associated with spin-1/2, and $\bar{\psi} \equiv \psi^\dagger\gamma^0$, but we will not
need to be very concerned about such details. What is
important is that $\mathcal{L}_e$ is not invariant under local transformations, but just as in quantum mechanics local invariance
can be imposed by the minimal substitution $\partial^\mu \to D^\mu \equiv$
$\partial^\mu - ieA^\mu$, where D^μ is the gauge covariant derivative.

---

[5]The Dirac matrices satisfy $\gamma^\mu\gamma^\nu + \gamma^\nu\gamma^\mu = g^{\mu\nu}$ and $\gamma^{\mu\dagger} = \gamma_\mu$. Their
exact representative is not needed here.

The QED Lagrangian density

$$\mathcal{L} = -\frac{1}{4}F_{\mu\nu}F^{\mu\nu} + \bar{\psi}\left[\gamma^{\mu}(i\partial_{\mu} + eA_{\mu}) - m_{e}\right]\psi$$
(4.16)

is invariant under the simultaneous gauge transformations

$$A'^{\mu} = A^{\mu} - \frac{1}{e}\partial^{\mu}\beta, \qquad \psi' = e^{-i\beta}\psi, \qquad (4.17)$$

i.e., $\mathcal{L}(\psi, A^{\mu}) = \mathcal{L}(\psi', A'^{\mu})$. This has several consequences:

- The photon must (at least apparently) be massless, because the addition of an elementary mass term $\frac{1}{2}m_{\gamma}A^{\mu}A_{\mu}$ to $\mathcal{L}$ would break the gauge invariance and, it turns out, the renormalizability.

- The $U(1)$ phase invariance, even at the global level, implies that the total lepton number L, i.e., the number of electrons minus the number of positrons, is conserved in all reactions.[6] L conservation is equivalent to the continuity equation $\partial_{\mu}J^{\mu} = 0$, where e times the electromagnetic current operator $J^{\mu} \equiv -\bar{\psi}\gamma^{\mu}\psi$ is the QED analog of $(\rho, \vec{J})$.

- The amplitude for the emission or absorption of a photon or for the creation or annihilation of an $e^{-}e^{+}$ pair is specified, up to the charge e, by the $e\,\bar{\psi}\gamma^{\mu}\psi A_{\mu} = -eJ^{\mu}A_{\mu}$ term in (4.17), multiplied by i, as is illustrated in figure 4.1. To see this in more detail, consider the schematic expressions in

---

[6]This is an example of the *Noether theorem*, associated with any continuous symmetry in classical or quantum mechanics.

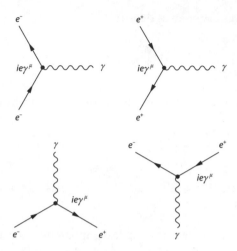

Figure 4.1. Vertices describing basic QED processes, from Langacker 2010. The particles may be real or virtual, and we have taken the convention that the initial (final) state is at the bottom (top) of the diagram. The photons in the first two vertices may be either incoming or outgoing. The arrow on the fermion line indicates the direction of flow of negative charge, i.e., along the momentum for an e^- or opposite the momentum direction for an e^+. There are additional factors, not shown, involving the $e^\pm$ spin and the photon polarization.

(4.13) and (4.14). Omitting the Lorentz, spin, and momentum details,

$$\psi^\dagger \psi A \sim (b^\dagger + d)(a + a^\dagger)(b + d^\dagger)$$
$$\sim b^\dagger a^\dagger b - d^\dagger a^\dagger d + a^\dagger db + b^\dagger d^\dagger a,$$
$$(4.18)$$

where $dd^\dagger = -d^\dagger d$ from Fermi statistics. The four terms correspond to the four vertices in figure 4.1. (We have taken the photon to be outgoing in the first two terms.)

- The electric charge of the photon is $q_\gamma = 0$, i.e., it has no elementary self-interactions.

The amplitudes for physical processes such as $e^- e^+ \to e^- e^+$ (Bhabha scattering) or $\gamma e^- \to \gamma e^-$ (Compton scattering) may be obtained from the interaction vertices in figure 4.1 using the rules of time-dependent perturbation theory. Feynman diagrams are an especially convenient version in which each term in the expansion can be represented by a diagram such as in figure 4.2, with specific factors associated with each vertex and each external or internal line. In this formalism, four-momentum is conserved at each vertex but the intermediate particles are virtual. For example, the Bhabha scattering amplitude is proportional to the matrix element

$$\langle e^-(p_3) e^+(p_4)| \left(e\, J^\mu A_\mu \right) \left(e\, J^\nu A_\nu \right) |e^-(p_1) e^+(p_2) \rangle, \tag{4.19}$$

where the p_i are the four-momenta of the external particles. This schematically reduces to two terms, corresponding to the two Feynman diagrams in figure 4.2. The first is

$$e^2 \langle e^- e^+ | b^\dagger d^\dagger |0\rangle \, \langle 0| a a^\dagger |0\rangle \, \langle 0| b d |e^- e^+ \rangle, \tag{4.20}$$

where $|0\rangle$ is the ground state (vacuum). The creation and annihilation operators in the fermion fields act on the external states, while those of the two A fields can be combined to form a *propagator* $\propto g_{\mu\nu}/q^2$, where q is the virtual four-momentum carried by the photon. The propagator in the second diagram in figure 4.2 leads to the long-range

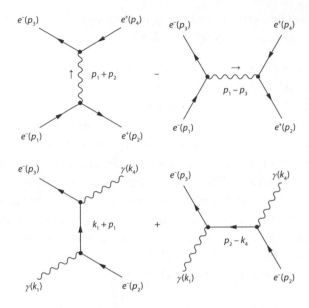

Figure 4.2. Lowest-order (tree-level) Feynman diagrams for $e^-e^+ \to e^-e^+$ and $\gamma e^- \to \gamma e^-$, from Langacker 2010. Each vertex is proportional to e, while each external or internal line has factors related to spin and polarization. Four-momentum is conserved at the vertices. In these examples, each process has two diagrams because of different ways of associating the external particles with the fields; the relative minus sign for $e^-e^+ \to e^-e^+$ is due to Fermi statistics.

potential $\propto e^2/r$ in position space. Similarly, the virtual electron lines for Compton scattering are proportional to $1/(q^2 - m_e^2)$, corresponding to $e^2 \exp(-m_e r)/r$.

QED is arguably the most successful theory in the history of science if judged by its quantitative results. For example, the magnetic moment of the electron is given by $\vec{\mu} = -g_e \mu_B \vec{S}$, where $\mu_B = e/2m_e$ is the Bohr magneton

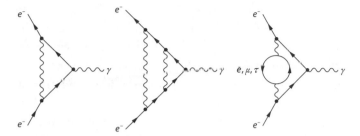

Figure 4.3. One- and typical two-loop diagrams contributing to the anomalous magnetic moment of the electron, from Langacker 2010.

and $\vec{S}$ is the electron spin operator. The "g-factor" g_e is not predicted by nonrelativistic quantum mechanics, but was known to be close to 2 from the Zeeman effect. The Dirac equation predicts $g_e = 2$, while there are QED corrections that can be computed as a power series in α/π. These are due to modifications of the electromagnetic vertex from diagrams[7] such as those in figure 4.3. The leading (one-loop) contribution, calculated by Julian Schwinger in 1947, yields the *anomalous magnetic moment* $a_e \equiv (g_e - 2)/2 = \alpha/2\pi \sim 0.00116$, where $\alpha = e^2/4\pi \sim 1/137$ is the fine structure constant. This was in excellent agreement with the experimental value at the time.

Currently, a_e has been calculated to the five-loop level, $\propto (\alpha/\pi)^5$, including the effects of hadrons and of weak interactions (Aoyama et al. 2015), and has been measured at Harvard with a precision of 0.24 ppb (0.24×10^{-9})

---

[7]In such diagrams involving closed loops of virtual particles, some of the internal momenta are not fixed by momentum conservation and must be integrated over. In a generic field theory, these integrals are divergent, but in renormalizable theories such as QED the final results are always finite.

using electrons held in a Penning trap. This is not by itself a test of QED, but rather can be used as a precise determination of α. A completely independent determination is obtained from the recoil velocity of the ^{87}Rb atom after emitting or absorbing a photon[8] (Bouchendira et al. 2011). The two values are $\alpha_{a_e}^{-1} = 137.035999157(33)$ and $\alpha_{Rb}^{-1} = 137.035999049(90)$, where the numbers in parentheses are the uncertainties in the last digits. These are in excellent agreement, confirming QED at the $\lesssim 10^{-9}$ level. Many other results in atomic physics (Lamb shift, hyperfine splittings), muonium ($\mu^+ e^-$) and positronium $e^+ e^-$, and the quantum Hall effect are also in agreement. See Karshenboim 2005 and Mohr et al. 2012 for reviews.

Two of the fundamental predictions of QED are that the photon should be massless and that it does not itself carry electric charge. Both are impressively confirmed by astrophysical observations (see Olive et al. 2014): $m_\gamma < 10^{-18}$ eV (from the Solar magnetic field) and $q_\gamma < 10^{-35}\, e$ from the isotropy of the CMB.

Despite these successes, there are two deviations from the predictions of QED that are not understood. One involves the anomalous magnetic moment a_μ of the muon, which has been measured to better that one part per million in a storage ring at the Brookhaven National Laboratory (BNL). a_μ has also been calculated through five loops, but is some 3.6σ lower than the observation (see Olive et al. 2014): $a_\mu^{exp} - a_\mu^{th} = 288(63)(49) \times 10^{-11}$,

---

[8]The experiment measures h/m_{Rb}, which can be combined with the Rydberg constant and mass ratios to obtain α.

where the two uncertainties are respectively experimental and theoretical. a_μ receives much larger weak interaction and hadronic contributions than a_e, and it is conceivable that the discrepancy is due to underestimated hadronic uncertainties. However, a_μ is also much more sensitive [by at least $(m_\mu/m_e)^2$] to new physics effects, such as diagrams involving new heavy particles. A more precise measurement of a_μ is in preparation at Fermilab.

The second discrepancy involves the Lamb shift (the $2S_{1/2} - 2P_{1/2}$ energy difference) in muonic ($\mu^- p$) hydrogen, which has been recently measured precisely at the Paul Scherrer Institut (PSI) in Switzerland. The Lamb shift is sensitive to the finite size of the proton, and the result is often expressed in terms of a derived radius. The muonic value for the proton radius is 0.84087(30) fm, some 7σ lower than the value 0.8775(51) fm obtained from ep scattering and the spectroscopy of ordinary atoms (Mohr et al. 2012). This could possibly be due to new physics, but (as of this writing) no one has found a very plausible model (Carlson 2015).

4.2 Internal Symmetries

A *symmetry* in physics refers to a transformation of a system onto a new one with identical properties (e.g., Barnes 2010). For example, rotations in a rotationally invariant potential map a classical or quantum system onto another with the same energy and properties. Three-dimensional rotations are described by $SU(2)$, which is the group of 2×2 unitary matrices with unit determinant.[9] The energy eigenstates of an invariant Hamiltonian fall into degenerate multiplets of dimension $2j + 1$, where $j = 0, \frac{1}{2}, 1, \frac{3}{2}, \cdots$. Under rotations, each of these states is transformed into a linear combination of the states in the multiplet.

The rotations are an example of a *space-time* symmetry. Other examples are Lorentz transformations, space-time translations, space reflection, and time reversal. Internal symmetries, on the other hand, involve the particles and their intrinsic properties themselves. For example, isospin refers to the approximate ($\sim$1%) symmetry of the strong interactions relating the proton and neutron; the three pions $\pi^{\pm}, \pi^0$; the $\mathcal{S} = 1$ kaons K^+, K^0, etc., into nearly degenerate multiplets (with mass splittings due to electroweak interactions and small quark mass differences). Isospin is mathematically identical to rotational symmetry. The transformations are described by the same $SU(2)$ group, though the physical interpretation is different. Particle multiplets involve $2I + 1$ states that can be "rotated" into each other, where the total isospin is $I = 0, \frac{1}{2}, 1, \frac{3}{2}, \cdots$.

For example, consider the isospin-invariant Lagrangian density for pions and nucleons. Let N be a two-component spinor containing the p and n fields, analogous to the spin orientations of a spin-1/2 particle, while $\vec{\pi}$ is a vector in the internal isospin space containing the three pion field

---

[9] $SU(2)$ is similar to the $SO(3)$ group of classical three-dimensional rotations, but allows half-integer spin.

components, analogous to a spin-1 rotational vector,

$$N = \begin{pmatrix} p \\ n \end{pmatrix}, \qquad \vec{\pi} = \begin{pmatrix} \pi_1 \\ \pi_2 \\ \pi_3 \end{pmatrix}, \qquad (4.21)$$

with $\pi^\pm = (\pi_1 \mp i\pi_2)/\sqrt{2}$ and $\pi^0 = \pi_3$. A global isospin rotation can be parametrized by three real numbers β_i, $i = 1, 2, 3$, just as an ordinary rotation involves three angles. For small β_i,

$$N \to \left(I + \frac{i}{2} \vec{\beta} \cdot \vec{\tau} \right) N, \qquad \pi_i \to \pi_i - \epsilon_{ijk} \beta_j \pi_k.$$

$$(4.22)$$

I is the 2×2 identity matrix and $\vec{\beta} \cdot \vec{\tau} \equiv \sum_{i=1}^{3} \beta_i \tau_i \equiv \beta_i \tau_i$, where the τ_i are the three Pauli matrices,[10] listed in table 4.1, which satisfy the commutation rules

$$\tau_i \tau_j - \tau_j \tau_i = 2i\epsilon_{ijk}\tau_k. \qquad (4.23)$$

The *structure constants* ϵ_{ijk} in (4.22) and (4.23) are the totally antisymmetric (Levi-Civita) tensor with $\epsilon_{123} = 1$. Sums over repeated indices are implied.

It is then straightforward to show that the Lagrangian density

$$\mathcal{L}_{\pi N} = \overline{N}\left(i\gamma^\mu \partial_\mu - m_N + ig_\pi \gamma^5 \vec{\pi} \cdot \vec{\tau}\right) N$$

$$+ \frac{1}{2}\left[\left(\partial_\mu \vec{\pi}\right)^2 - m_\pi^2 \vec{\pi}^{\,2}\right] - \lambda(\vec{\pi}^{\,2})^2 \quad (4.24)$$

---

[10] These are usually denoted σ_i when discussing spin.

Table 4.1. The $SU(2)$ Pauli matrices τ_i, the $SU(3)$ Gell-Mann matrices λ_i, and the nonzero $SU(3)$ structure constants f_{ijk}.

$$\tau_1 = \begin{pmatrix} 0 & 1 \\ 1 & 0 \end{pmatrix} \qquad \tau_2 = \begin{pmatrix} 0 & -i \\ i & 0 \end{pmatrix} \qquad \tau_3 = \begin{pmatrix} 1 & 0 \\ 0 & -1 \end{pmatrix}$$

$$\lambda_{1,2,3} = \begin{pmatrix} & & 0 \\ \tau_{1,2,3} & & 0 \\ 0\ 0 & & 0 \end{pmatrix} \quad \lambda_4 = \begin{pmatrix} 0 & 0 & 1 \\ 0 & 0 & 0 \\ 1 & 0 & 0 \end{pmatrix} \quad \lambda_5 = \begin{pmatrix} 0 & 0 & -i \\ 0 & 0 & 0 \\ i & 0 & 0 \end{pmatrix}$$

$$\lambda_6 = \begin{pmatrix} 0 & 0 & 0 \\ 0 & 0 & 1 \\ 0 & 1 & 0 \end{pmatrix} \quad \lambda_7 = \begin{pmatrix} 0 & 0 & 0 \\ 0 & 0 & -i \\ 0 & i & 0 \end{pmatrix} \quad \lambda_8 = \frac{1}{\sqrt{3}} \begin{pmatrix} 1 & 0 & 0 \\ 0 & 1 & 0 \\ 0 & 0 & -2 \end{pmatrix}$$

$f_{123} = 1$	$f_{147} = \frac{1}{2}$	$f_{156} = -\frac{1}{2}$
$f_{246} = \frac{1}{2}$	$f_{257} = \frac{1}{2}$	$f_{345} = \frac{1}{2}$
$f_{367} = -\frac{1}{2}$	$f_{458} = \frac{\sqrt{3}}{2}$	$f_{678} = \frac{\sqrt{3}}{2}$

is isospin invariant. m_N and m_π are respectively the common mass for the two nucleons and for the three pions, g_π is the coupling constant, γ^5 is a Dirac matrix[11] associated with the pseudoscalar nature of the pion, and λ is the strength of the pion self-interaction. The πN

---

[11] N has two sets of indices. Each of the two isospin components p and n also has a suppressed Dirac index running from 1 to 4. The 2×2–dimensional $\vec{\tau}$ matrices in (4.24) act on the isospin index, while the 4×4–dimensional γ^5 acts on the Dirac index.

interaction term in (4.24) can be rewritten

$$ig_\pi \overline{N}\gamma^5 \vec{\pi} \cdot \vec{\tau} N = ig_\pi \left[(\bar{p}\gamma^5 p - \bar{n}\gamma^5 n)\pi^0 \right.$$
$$\left. + \sqrt{2}(\bar{p}\gamma^5 n\pi^+ + \bar{n}\gamma^5 p\pi^-) \right],$$
$$(4.25)$$

illustrating how the isospin symmetry relates the various couplings. $\vec{\pi} \cdot \vec{\tau}$ is analogous to the spin-orbit coupling $\propto \vec{L} \cdot \vec{\sigma}$ in the hydrogen atom.

At a more fundamental level, isospin is a (flavor) symmetry between the u and d quarks, which transform as an isospin doublet. Their QCD interactions are identical because the gluons are isospin singlets. Their (Higgs-induced) Langrangian masses in table 3.2 are different, but both are tiny compared to QCD-induced masses so that isospin breaking is small.[12] As mentioned in chapter 2, Gell-Mann and Ne'eman extended isospin to a larger eightfold way symmetry between strange and nonstrange particles. This is based on the $SU(3)$ group of 3×3 unitary matrices with unit determinant, and can be understood as a flavor symmetry between the u, d, and s quarks, which transform as a three-dimensional (triplet) representation. The s quark is considerably more massive ($m_s \sim 95$ MeV), so that the mass splittings in the hadron multiplets are around 25%. However, the symmetry works much better for interaction strengths, which can again be understood because the QCD interactions are the same for all flavors. The c, b, and t quarks are much heavier than the

---

[12]Electromagnetism by itself would make the proton heavier than the neutron. However, the quark mass effect ($m_d > m_u$) is slightly larger, making the neutron heavier. Similar statements apply to K^+ and K^0. However, the quark masses don't affect $m_{\pi^+} - m_{\pi^0}$ to leading order, so the π^+ is heavier.

others, so the extension to still higher flavor symmetries is not very useful.

4.3 Yang-Mills Theories

Both $SU(2)$ (isospin) and $SU(3)$ (eightfold way) are *nonabelian* symmetries. This means that the transformations do not in general commute, in contrast with the $U(1)$ group of phase factors discussed earlier. They are also global, and can be explicitly broken by quark mass differences and electroweak interactions without disastrous consequences.

In 1954, C. N. Yang and Robert Mills showed that $U(1)$ gauge invariance can be generalized to higher non-abelian symmetries like $SU(2)$ and $SU(3)$ (Yang and Mills 1954). When spin-1 mesons such as the ρ in table 3.2 were later discovered, it was thought that these might be the Yang-Mills gauge bosons, e.g., extending isospin to a gauge symmetry. This never worked out, in part because of the ρ masses, but the mathematics of $SU(2)$ and $SU(3)$ was recycled in the standard model, in which the electroweak and strong interactions are associated with non-abelian gauge symmetries distinct from and in addition to the global flavor symmetries of the strong interactions.

Let us consider a hypothetical $SU(2)$ gauge symmetry in which two fermion fields ψ_1 and ψ_2 transform as a doublet, i.e., are rotated into each other,

$$\psi \rightarrow \exp\left[\frac{i}{2} \vec{\beta}(x) \cdot \vec{\tau}\right] \psi \sim \left[I + \frac{i}{2} \vec{\beta}(x) \cdot \vec{\tau}\right] \psi,$$
(4.26)

where $\psi \equiv \begin{pmatrix} \psi_1 \\ \psi_2 \end{pmatrix}$. Equation (4.26) generalizes[13] the $U(1)$ gauge transformation in (4.17) on page 59. We will always work with the second form of (4.26), valid for small $\vec{\beta}$. Gauge invariance requires that momenta are replaced by the minimal substitution $\partial^\mu \to D^\mu \equiv \partial^\mu + ig\,\vec{A}^\mu \cdot \vec{\tau}/2$, where the $A_i^\mu(x)$, $i = 1, 2, 3$ are three gauge bosons corresponding to the three $\beta_i(x)$, and g is a gauge coupling constant. Under a gauge transformation,

$$A_i^\mu \to A_i^\mu - \epsilon_{ijk}\,\beta_j\,A_k^\mu - \frac{1}{g}\,\partial^\mu \beta_i(x). \qquad (4.27)$$

The second term on the right reflects the fact that, unlike QED, the gauge bosons transform nontrivially (as spin-1) even under global transformations. They are analogous to the pion transformation in (4.22) for an isospin rotation. The last term is analogous to (4.17).

The nontrivial global transformation of the gauge fields requires that the field strength tensor must be modified from (4.10) to

$$F_i^{\mu\nu} = \partial^\mu A_i^\nu - \partial^\nu A_i^\mu - g\,\epsilon_{ijk}\,A_j^\mu\,A_k^\nu. \qquad (4.28)$$

The $SU(2)$ gauge invariant Lagrangian density is

$$\mathcal{L} = -\frac{1}{4}F_{i\,\mu\nu}F_i^{\mu\nu} + \bar{\psi}\left[\gamma^\mu\left(i\partial_\mu - \frac{g}{2}\,\vec{A}_\mu \cdot \vec{\tau}\right) - m\right]\psi,$$
$$(4.29)$$

---

[13]The extra minus sign in (4.17) is because the electron charge is $-e$, due to an unfortunate sign convention made by Benjamin Franklin.

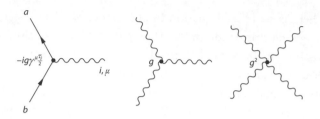

Figure 4.4. $SU(2)$ interaction vertices corresponding to (4.29). The rather complicated dependence of the gauge self-interactions on the $SU(2)$ and Lorentz indices can be determined from (4.28).

where m is the common fermion mass. No elementary mass term for the gauge fields is allowed. Not only are there vertices for the emission or absorption of gauge bosons by the fermions, but there are now also three and four-point self-interactions amongst the gauge bosons, as illustrated in figure 4.4. Once the matter content has been specified [one $SU(2)$ doublet of fermions] the interactions are uniquely determined by gauge invariance up to one coupling constant g. Such theories are mathematically well-behaved (renormalizable) and are the unique field theories involving spin-1 particles (in four space-time dimensions) with this property.

The $SU(3)$ group plays the dual roles of an approximate global symmetry acting on the (u, d, s) quark flavor indices and as a gauge theory (QCD) acting on the quark color indices. The structure of an $SU(3)$ gauge theory can be easily generalized from $SU(2)$, except that now the lowest-dimensional nontrivial multiplet is a triplet and there are eight analogs of the three $SU(2)$ rotation angles. Equations (4.28) and (4.29) are still valid after making the following substitutions:

- Replace ψ in (4.26) by a triplet (ψ_1, ψ_2, ψ_3), e.g., the three colors of the up quark.
- Let i run from 1 to 8 in β_i and A_i^μ.
- Generalize the Pauli matrices τ_i to the eight 3×3–dimensional *Gell-Mann matrices* λ_i listed in table 4.1.
- Replace ϵ_{ijk} by the totally antisymmetric f_{ijk}, whose nonzero values are listed in table 4.1.

4.4 Quantum Chromodynamics

In sections 4.1 and 4.3, we explored the structure of $U(1)$, $SU(2)$, and $SU(3)$ gauge theories. All three are utilized in the standard model, based on the group of transformations $SU(3) \times SU(2) \times U(1)$. The notation means that the three parts of the group commute with each other and can have different coupling constants. We start with quantum chromodynamics (QCD), which describes the strong interactions of quarks and gluons. QCD was developed in the early 1970s. It actually postdated the electroweak theory, but is less complicated. Detailed discussions of the history are given in Gross 2005; Leutwyler 2014; and Fritzsch 2014; and the current status in Olive et al. 2014.

The Ingredients

QCD is a gauge theory based on $SU(3)$. The coupling constant and eight gluons are usually written as g_s and G_i^μ

Chapter 4

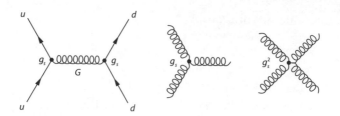

Figure 4.5. The basic QCD interactions. The squiggly lines represent gluons, from Langacker 2010.

rather than the generic labels in section 4.3. Thus,

$$
\begin{aligned}
\mathcal{L}_{QCD} = {}& -\tfrac{1}{4} G_{i\mu\nu} G_i^{\mu\nu} \\
& + \bar{u} \left[\gamma^\mu \left(i \partial_\mu - g_s G_{i\mu} \frac{\lambda_i}{2} \right) - m_u \right] u + \cdots,
\end{aligned}
\tag{4.30}
$$

where a sum on i from 1 to 8 is implied and

$$
G_i^{\mu\nu} = \partial^\mu G_i^\nu - \partial^\nu G_i^\mu - g_s f_{ijk} G_j^\mu G_k^\nu. \tag{4.31}
$$

u is a spinor with components u_r, u_g, and u_b and the dots represent terms for the d, c, s, t, and b quarks.

The $SU(3)$ interactions change the quark color but not the flavor, e.g., there are transitions between u_r, u_g, and u_b with the emission or absorption of a gluon (or diagonal couplings such as $u_r \to u_r$ for G_3 and G_8). The amplitude for these transitions is $-ig_s\lambda_i/2$. These couplings are the same for each flavor and for both left- and right-chiral quarks (i.e., parity is conserved). The leptons are singlets, i.e., they are not affected by $SU(3)$ transformations. The basic QCD interactions are illustrated in figure 4.5.

The original evidence for the existence of quarks came from hadron spectroscopy. Also, the global hadronic flavor symmetries (isospin, the eightfold way, and their extension to chiral symmetries) and the patterns of their breaking emerged as rather natural consequences. However, most physicists were skeptical about their actual existence because no isolated fractionally-charged state was directly observed. Most doubts were answered by the deep inelastic scattering (DIS) experiments at the Stanford Linear Accelerator Center (SLAC), which observed the reaction $e^- N \rightarrow e^- X$ via the exchange of a virtual photon (figure 3.1), where N is a nucleus and X represents a sum over final hadronic states, typically consisting of a large number of pions and nucleons. Surprisingly, the cross section was much larger for large transfers of momentum from the electron to the hadrons than would be expected if the nucleon were a big continuous object of size $\sim 10^{-13}$ cm. This suggested that it is really a bound state of much smaller or point-like constituents. The detailed angular and energy dependence implied that these had spin-1/2, consistent with quarks.[14] Apparently the quarks really exist, but for some reason they are confined (cannot emerge as free particles), but rather radiate their energy away in a somewhat collimated *jet* of hadrons.

Despite its successes, the original quark model had another major difficulty: the baryon spectrum implies that the three-quark wave functions are totally symmetric under the exchange of spin, space, and flavor indices, in violation

[14]This is reminiscent of the Rutherford experiment that established the existence of the atomic nucleus, only scaled up in energy by a thousand.

of the spin-statistics theorem. This is most evident in the Ω^-, the particle that was successfully predicted by the eightfold way as mentioned in section 2.3. The Ω^- consists of three identical s quarks, with the spins in a symmetric $S = 3/2$ combination and a symmetric space wave function with orbital angular momentum $L = 0$. Fermi statistics is nicely restored by the existence of color because the the three quarks in a baryon are in an anti-symmetric color-singlet state. More direct evidence that there are really three colors for each flavor emerged later from various counting experiments and observations. For example, the reaction $e^- e^+ \rightarrow$ hadrons is really due to $e^- e^+ \rightarrow q\bar{q}$ via a virtual photon, analogous to the first diagram in figure 4.2 except the final $e^- e^+$ pair is replaced by $q\bar{q}$. As in DIS, the final q and $\bar{q}$ emerge as jets of hadrons. The total rate is obtained by summing over the quarks (with appropriate electric charges) and is therefore proportional to the number N_c of colors. Experiments at SLAC and elsewhere clearly established $N_c = 3$ and not one, as can be seen in figure 4.6.

The first direct evidence for spin-1 gluons also came from $e^- e^+ \rightarrow$ hadrons. Occasionally, the final q or $\bar{q}$ can radiate a gluon, leading to three jets rather than two, or to a broadening of a jet if the third one is not resolved. This was first observed at the Deutsches Elektronen-Synchrotron (DESY) laboratory in Germany in 1979. Apparently, gluons are also confined.

The Long and the Short

The strong interactions manifest themselves in two very different ways, depending on the circumstances. In the

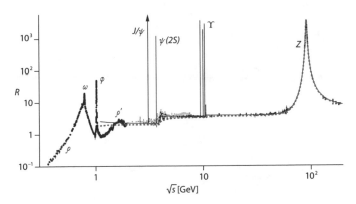

Figure 4.6. Cross section for $e^+e^- \to$ hadrons relative to $e^+e^- \to \mu^+\mu^-$ versus the CM energy, from Olive et al. 2014. The low-energy (photon-dominated) region verifies that there are three colors and probes the QCD corrections. The peaks are from the production of spin-1 $\bar{q}q$ mesons, including $\bar{c}c$ ($J/\psi, \psi$) and $\bar{b}b$ (Υ). The Z peak tests the electroweak theory, and the intermediate region involves $\gamma + Z$ interference.

low-energy ($\lesssim$ GeV) and long-distance ($\gtrsim$ fm) regime relevant to nuclear physics, the key players are the hadrons, the interaction strength is large, and quarks and gluons cannot be produced as isolated particles. At high energies and short distances, however, the fundamental degrees of freedom are quarks and gluons, and the interaction is actually rather weak. To first approximation deep inelastic scattering and $e^+e^- \to$ hadrons at high energy can be described by ignoring the strong interactions altogether in the underlying process, with the quarks and gluons turning into jets of hadrons much later. These two regimes, referred to as *infrared slavery* and *asymptotic freedom*, respectively, can be understood in QCD to be a consequence

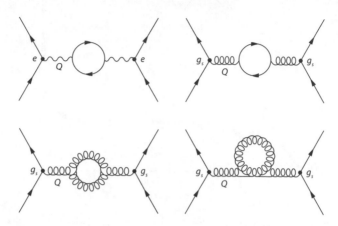

Figure 4.7. Diagrams contributing to the running of α in QED and α_s in QCD, from Langacker 2010.

of the *running* coupling constants. Interaction parameters such as gauge couplings in field theory are not actually constants. Rather, they depend on the typical momentum scale Q of the particles in the interaction.

To understand this, consider $e^- e^-$ scattering in QED. To lowest order, the scattering is due to a diagram analogous to the ep diagram in figure 3.1 on page 35 (and a second diagram in which the two final electrons are interchanged), which is proportional to $\alpha = e^2/4\pi$. However, there are higher-order corrections. The one relevant to the present discussion is the upper-left diagram in figure 4.7, in which the virtual photon temporarily turns into a virtual $e^+ e^-$ (or other charged particle) pair. This diagram causes the effective strength of the interaction to increase logarithmically in the momentum Q carried by the photon, or equivalently to fall more rapidly than

$1/r$ as r increases. This *vacuum polarization* is the analog of charge screening in a dielectric medium, with virtual particles playing the role of the dielectric, and the effect can be absorbed into a running (or effective) $\alpha(Q^2)$. This running is small for QED: the value $\alpha \sim 1/137$ relevant at low energies increases to $\alpha(M_Z^2) \sim 1/128$ at the scale of Z-pole physics to be described in section 4.7, but it is observed and is actually quite important for the precision tests.

The running is even more important in QCD. The virtual quark pair diagrams in figure 4.7 by themselves would also cause the strong fine structure constant $\alpha_s \equiv g_s^2/4\pi$ to increase at higher energies. However, the vitural gluon diagrams anti-screen, and are larger (for no more than 16 quark flavors). Thus α_s decreases for large Q^2 (asymptotic freedom). The running has been verified by extracting α_s from a variety of processes at different Q^2 values and comparing with the QCD prediction, as can be seen in figure 4.8. At a high enough scale, $\alpha_s(Q^2)$ is sufficiently small that the quarks act as nearly free particles, with QCD corrections calculable in a power series in $\alpha_s(Q^2)/\pi$. Conversely, α_s becomes large for small Q (infrared slavery). At some scale $\Lambda_{QCD} \lesssim 1\,\text{GeV}$ (which is determined by fitting to the data), α_s becomes so large that perturbation theory breaks down. The quarks acquire masses $\sim m_p/3$ from the gluon clouds surrounding them (in addition to the Higgs-induced masses) and bind into hadrons.

It is interesting that, in the absence of the Higgs-induced quark masses, QCD would have no free parameters because the interaction strength α_s runs from large

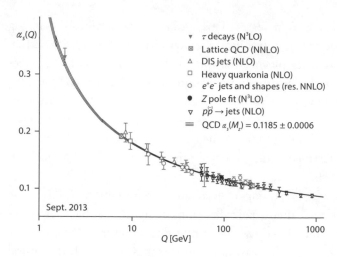

Figure 4.8. $\alpha_s(Q^2)$ from various observations, compared to the QCD fit, from Olive et al. 2014.

values to zero at different scales. Λ_{QCD}, which controls the running, could simply be *defined* as the scale for which, e.g., $\alpha_s = 1$. Only dimensionless ratios like m_p/Λ_{QCD}, which would in principle be calculable, would be physical. In practice, however, the quark masses and other non-QCD scales are present, so Λ_{QCD} must be determined by experiment.

Quark (and gluon) confinement can be understood in this picture, at least in a hand-waving way, because of the strong forces that develop as they are separated and because of the gluon self-interactions. Consider pulling apart a $q\bar{q}$ pair, for example. If there were no gluon self-interactions and the coupling were small, the gluon field would develop lines of flux that would spread out similarly to the electric field lines between an e^+e^- pair. However, the large self-interactions would instead cause the QCD flux lines to contract into a relatively narrow tube separating the $q\bar{q}$ pair. The energy stored in this flux tube would increase with its length, and for a large enough separation it would become energetically favorable to create a $\bar{q}q$ pair. The flux tube would break, with the created $\bar{q}$ and q at the ends. Instead of isolating the original q and $\bar{q}$, one would have two $q\bar{q}$ pairs (two mesons), just as cutting a bar magnet in half produces two bar magnets rather than a magnetic monopole and antimonopole. Color confinement can be established much more rigorously by lattice calculations, in which space and time are discretized so that the QCD equations of motion can be solved on a supercomputer.

Testing QCD

QCD has by now been verified in many ways. Perhaps the most important is the success of the running coupling predictions in terms of one fitted parameter, Λ_{QCD}, or alternatively $\alpha_s(M_Z^2)$, as can be seen in figure 4.8. Nonabelian gauge theories are the only renormalizable field theories that exhibit asymptotic freedom in four dimensions (Gross and Wilczek 1973; Politzer 1973), and QCD is essentially the unique possible theory incorporating the basic ideas of quarks with three colors.

There have been many successful tests of the quantitative predictions for short-distance (large Q^2) processes. These have included deep inelastic scattering (culminating in the high energy ep collider HERA at DESY) and

$e^+e^- \rightarrow$ hadrons. In both of these, the leading contribution is independent of α_s but the higher-order corrections are nonnegligible. Also, pp and $\bar{p}p$ scattering at high energies can be viewed as scattering of the constitutent quarks and gluons (followed by the later formation of hadron jets). Many final states, involving not only hadron jets but also electroweak particles in the final states have been measured at the Tevatron at Fermilab and at the LHC at CERN, with excellent agreement with the expectations of QCD and the electroweak theory. Typical jet results are shown in figure 4.9.

The long-distance regime has also been extensively tested, in the spectroscopy, decays, and weak interactions of both the ordinary hadrons and those involving heavy b and c quarks, as well as the flavor symmetries of the lighter hadrons. The lattice calculations have been especially powerful theoretically, not only in predicting most of the hadron masses from a few inputs, but also in calculating strong interaction matrix elements relevant to weak interaction decays.

To summarize, the strong interactions, once believed by many to be hopelessly complicated, are in fact well described by quantum chromodynamics. QCD is well behaved mathematically, and in the absence of other scales from gravity or the electroweak sector, it would have no free parameters. Asymptotic freedom implies that QCD would make sense at arbitrarily high energies (even in the absence of new physics scales such as M_P), and has the bonus that sensible calculations can be carried out concerning its behavior at very high temperatures in the early Universe.

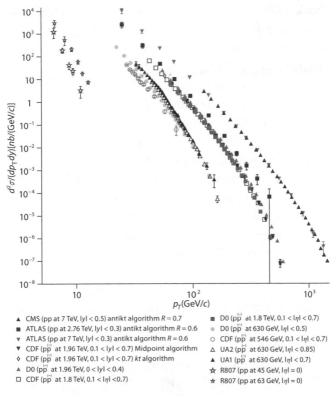

Figure 4.9. Production cross sections for jets in pp and $\bar{p}p$ interactions as a function of momentum transverse to the beam direction for a wide range of beam energies, from Olive et al. 2014. The results are in excellent agreement with the predictions of QCD. (Some of the lower-energy data was used in the determination of the quark and gluon distributions in the proton.)

4.5 The $SU(2) \times U(1)$ Model

The original $SU(2) \times U(1)$ model unifying the weak and and electromagnetic interactions was proposed in 1961 by

Sheldon Glashow, although he did not have a satisfactory
mechanism for generating masses for the gauge bosons
and chiral fermions. A more complete theory for leptons
that utilized the Higgs mechanism for the masses was
given by Steven Weinberg in 1967 and independently by
Abdus Salam the following year. Renormalizability was
established by Gerard 't Hooft and Martinus Veltman a
few years later, while the discovery of the charm c quark
in 1974 allowed a realistic extension to hadrons. The
$SU(2) \times U(1)$ model incorporated QED and the Fermi
theory of the weak charged current (WCC) interaction,
but improved on the latter by allowing sensible higher-
order corrections. It predicted the existence of a new weak
neutral current (WNC) interaction, of the $W^{\pm}$ and Z
gauge bosons, and of the Higgs boson. The development
of the $SU(2) \times U(1)$ model is described in detail in
Weinberg 1980a; Glashow 1980; and Salam 1980, while
the current status is reviewed in Langacker 2010 and Olive
et al. 2014.

$SU(2)$ has gauge coupling g and gauge bosons
$W_i^\mu, i = 1, 2, 3$. $W^{\pm\mu} \equiv (W_1^\mu \mp i W_2^\mu)/\sqrt{2}$ are emitted
or absorbed with strength $-ig/\sqrt{2}$ in the WCC
transitions between the members of each left-chiral
doublet,[15] $\begin{pmatrix} u_r \\ d_r \end{pmatrix}_L$, $\begin{pmatrix} u_g \\ d_g \end{pmatrix}_L$, $\begin{pmatrix} u_b \\ d_b \end{pmatrix}_L$, $\begin{pmatrix} \nu_e \\ e^- \end{pmatrix}_L$, and the two
heavier families, while $W^{0\mu} \equiv W_3^\mu$ couples diagonally
with strength $-ig\,\tau_3/2$. The right-chiral fermions are
all singlets and do not couple to the W_i^μ, and weak
transitions do not change the quark color.

---

[15] The mixing between families will be described later.

The $U(1)$ has coupling g' and gauge boson B^μ. It interacts diagonally with all fermions except the ν_R, but the strengths for left- and right-chiral are different. The coupling of the B to particle p_L or p_R is $-ig' y_{p_{L,R}}$, where the $U(1)$ charges $y_{p_{L,R}}$ are

$$y_{p_L} = q_p - t^3_{pL}, \qquad y_{p_R} = q_p. \qquad (4.32)$$

q_p is the electric charge of p (in units of e) and $t^3_{pL} = \tau_3/2 = \pm 1/2$ for the upper (lower) components of the left-chiral doublets. Thus,

$$y_{u_L} = y_{d_L} = \frac{1}{6}, \quad y_{\nu_L} = y_{e_L^-} = -\frac{1}{2},$$

$$y_{u_R} = \frac{2}{3}, \quad y_{d_R} = -\frac{1}{3}, \quad y_{\nu_R} = 0, \quad y_{e_R^-} = -1.$$
$$(4.33)$$

The reason for this rather nonintuitive charge assignment will become clear after applying the Higgs mechanism, when the γ and Z will be seen to be linear combinations (mixtures) of W^0 and B.

Let us put all of this together and display the interaction terms for the u and d quarks:

$$\begin{aligned}
\mathcal{L} = &- \frac{g}{2} \left[(\bar{u}_L \gamma^\mu u_L - \bar{d}_L \gamma^\mu d_L) \, W^0_\mu \right. \\
&\left. + \sqrt{2} \, \bar{u}_L \gamma^\mu d_L \, W^+_\mu + \sqrt{2} \, \bar{d}_L \gamma^\mu u_L \, W^-_\mu \right] \\
&- g' \left[\frac{1}{6} (\bar{u}_L \gamma^\mu u_L + \bar{d}_L \gamma^\mu d_L) \right. \\
&\left. + \frac{2}{3} \bar{u}_R \gamma^\mu u_R - \frac{1}{3} \bar{d}_R \gamma^\mu d_R \right] B_\mu, \qquad (4.34)
\end{aligned}$$

where a sum over the quark colors is implied. u_L and u_R are respectively the fields associated with the left (L)– and right (R)–chiral up quark.[16] Combinations such as $\bar{u}_L \gamma^\mu u_L$ and $\bar{u}_R \gamma^\mu u_R$ are known respectively as $V - A$ and $V + A$ currents, where vector (V) and axial vector currents (A) have opposite transformations under space reflection. Gauge theories such as $SU(2) \times U(1)$ with different couplings for left and right are known as chiral and are usually parity-violating.

The interactions for the heavier quarks are the same as in (4.34), while those for the leptons differ only in their $U(1)$ charges. As usual, the gauge invariance does not allow elementary mass terms for the W_i or B. Elementary fermion mass terms such as $-m_u \bar{u}u$, which would be fine for a non-chiral theory such as QCD or QED, are not invariant under a chiral transformation because they connect left and right, i.e., $\bar{u}u = \bar{u}_L u_R + \bar{u}_R u_L$.

4.6 The Higgs Mechanism

Most symmetries in nature are only approximate. A symmetry may be broken *explicitly* by a small perturbation, such as the breaking of rotational invariance when an atom is placed in an external magnetic field. Explicit breaking of a global symmetry in a quantum field theory is innocuous and in fact occurs for the flavor symmetries of the strong interactions. However, any explicit breaking of a gauge

---

[16]Technically, $u_{L,R} \equiv \frac{(1 \mp \gamma^5)}{2} u$, where the Dirac matrices project out the appropriate components of the Dirac field.

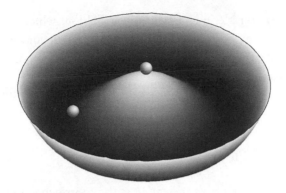

Figure 4.10. Wine bottle (or Mexican hat) potential. The rotationally symmetric point at the top is unstable. The lowest energy solution is at any point in the trough.

symmetry spoils the renormalizability and leads to severe mathematical difficulties.

Spontaneous symmetry breaking (SSB), in which the underlying symmetry is hidden, is another possibility. This means that the stable solutions exhibit less symmetry than the equations of motion. For example, the equations governing a ferromagnet are rotationally invariant, but the electron spins in a domain align in a definite direction. As another example, consider a marble moving inside an empty wine bottle, as illustrated in figure 4.10. The system has a rotational symmetry around the vertical axis, but the symmetric solution, with the marble at rest at the center, is unstable. The lowest-energy (stable) solution is for the marble to be at rest somewhere in the trough, obviously breaking the symmetry. There are also excitations with nonzero but arbitrarily small kinetic energy in which the marble rolls around the trough.

Now consider a complex spin-0 field ϕ, which (along with $\phi^\dagger$) can correspond to a charged particle and its antiparticle, such as $\pi^\pm$ or K^0, $\bar{K}^0$. The Lagrangian density

$$\mathcal{L}_\phi = (\partial^\mu \phi)^\dagger (\partial_\mu \phi) - V(\phi), \qquad (4.35)$$

where

$$V(\phi) \equiv \mu^2 \phi^\dagger \phi + \lambda (\phi^\dagger \phi)^2 \qquad (4.36)$$

is known as the *scalar potential*, is the most general renormalizable potential with a global $U(1)$ phase symmetry under $\phi \rightarrow e^{i\beta}\phi$. The corresponding conserved Noether charge is electric charge for the pion example, or strangeness for the neutral kaon.

The *vacuum* state $|0\rangle$ corresponding to (4.35) is not necessarily empty space. It is rather the state of lowest energy (i.e., the ground state), characterized by the *vacuum expectation value* (VEV) $\langle 0|\phi|0\rangle \equiv v/\sqrt{2}$, which can be thought of as a classical background field around which the theory is quantized. (The $\sqrt{2}$ is for later convenience.) v is the value of ϕ for which the potential $V(\phi)$ is minimized.[17] For $\mu^2 > 0$, the minimum[18] is for $v = 0$. In this case, the vacuum is indeed empty space, the quantum excitations for small λ correspond to particles or antiparticles of mass μ, and the Noether charge is conserved.

However, μ^2 is simply a parameter in the theory, which could very well be negative. For $\mu^2 < 0$, the point $v = 0$ is

[17]Nonzero derivatives $\partial_\mu \phi$ increase the energy, so v is independent of $\vec{x}$ and t.
[18]We assume that $\lambda \geq 0$. Otherwise, the potential is unbounded from below and there is no stable ground state.

an unstable maximum, and the minimum is at some value of $v \neq 0$.

Let us examine the $\mu^2 < 0$ case in more detail. It is convenient to write $\phi = (\phi_1 + i\phi_2)/\sqrt{2}$, where ϕ_1 and ϕ_2 are real (i.e., Hermitian). Then,

$$V = \frac{\mu^2}{2}\left(\phi_1^2 + \phi_2^2\right) + \frac{\lambda}{4}\left(\phi_1^2 + \phi_2^2\right)^2, \qquad (4.37)$$

which is similar to the wine-bottle potential in figure 4.10. It exhibits a one-dimensional rotational symmetry, which is just the $U(1)$ phase invariance expressed in the real components. There is a line of degenerate minima along the trough with $v \equiv (v_1^2 + v_2^2)^{1/2} = \sqrt{-\mu^2/\lambda}$, where $v_i = \langle 0|\phi_i|0\rangle$. Without loss of generality, we can choose $v_1 = v$ and $v_2 = 0$, so that the vacuum itself breaks the rotational invariance.

To quantize around the vacuum, we introduce "normal" fields with zero VEVs. It is convenient to use polar coordinates to make use of the rotational symmetry of the original potential, i.e.,

$$\phi = \frac{1}{\sqrt{2}}(v + H)\,e^{i\theta}, \qquad (4.38)$$

where H is a real field that represents vibrations in the radial direction and θ represents motion along the trough. The potential becomes

$$V = \frac{-\mu^4}{4\lambda} - \mu^2 H^2 + \lambda v H^3 + \frac{\lambda}{4}H^4, \qquad (4.39)$$

independent of θ. The symmetry implies that there is no minimum energy required for rolling around the potential, and the corresponding spin-0 particle is massless. The existence of such a massless particle, known as a *Nambu-Goldstone boson*, is an inevitable consequence of any spontaneously broken continuous global symmetry.[19]

The first term in (4.39) is a constant that has no physical consequence here. However, its analog will come back to haunt us when gravity is considered in chapter 5. The second term, the curvature in the radial direction at the minimum, represents a mass $\sqrt{-2\mu^2}$ for the physical spin-0 particle corresponding to the H field. The third and fourth terms are self-interactions. There is no conserved charge in this $\mu^2 < 0$ phase. For example, the $\lambda v H^3$ (*induced cubic*) term allows transitions between even and odd numbers of H particles.

Things become more interesting when the $U(1)$ is promoted to a gauge invariance. Using the minimal substitution, the kinetic term in (4.35) becomes

$$[(\partial^\mu + ig A^\mu)\phi]^\dagger(\partial_\mu + ig A_\mu)\phi \to \frac{(gv)^2}{2} A^\mu A_\mu,$$
$$(4.40)$$

where the second form is obtained when ϕ is replaced by its classical value $v/\sqrt{2}$. That is, the spontaneous breaking generates an effective mass $M_A = gv$ for the gauge field. This can be thought of as the effect of the A constantly interacting with the background field, analogous to the

---

[19]The small mass of the pions is a consequence of the small u and d quark masses. For $m_u = m_d = 0$, QCD would have a spontaneously broken chiral extension of isospin symmetry, and the pions would be massless Nambu-Goldstone bosons.

effective mass of a photon propagating through a plasma. This is the Higgs mechanism.[20] The important features are

- Unlike an elementary mass term, it does not spoil the renormalizability of the theory.
- A careful examination reveals that the massless Nambu-Goldstone boson θ disappears from the spectrum, reemerging as the longitudinal (helicity-0) polarization state of the massive vector boson. It is said to have been "eaten."
- There is a massive physical spin-0 H field (the analog of the Higgs boson). Its interactions with the gauge fields can be obtained from the first form of (4.40).

In the next section, the simple (and unrealistic) model described earlier will be extended to a realistic one, the standard electroweak theory.

4.7 The Electroweak Theory

The standard electroweak theory is obtained by applying the Higgs mechanism to spontaneously break the $SU(2) \times U(1)$ gauge symmetry so that three of the four gauge bosons acquire mass, as do the fermions. One linear combination, the photon, remains massless because the background Higgs field has no electric charge. The $U(1)$ of QED remains as an unbroken gauge invariance.

[20]The mechanism was proposed more or less at the same time by Peter Higgs; Robert Brout and François Englert; and Gerald Guralnik, Carl Hagen, and Tom Kibble. A nonrelativistic condensed matter analog had been given earlier by Philip Anderson. It was first applied to the weak interactions by Steven Weinberg. These developments are chronicled in detail in Quigg 2015.

The Gauge and Higgs Bosons

The simplest implementation of SSB for $SU(2) \times U(1)$ involves the introduction of a single Higgs doublet of spin-0 particles, $\phi = \begin{pmatrix} \phi^+ \\ \phi^0 \end{pmatrix}$. The $U(1)$ charge is $y_\phi = 1/2$, similar to that of the left-chiral fermions in (4.32). The Lagrangian density for ϕ is then

$$\mathcal{L}_\phi = (D^\mu \phi)^\dagger (D_\mu \phi) - V(\phi). \qquad (4.41)$$

The potential $V(\phi)$ still takes the form in (4.36), except now $\phi^\dagger \phi = \phi^{+\dagger} \phi^+ + \phi^{0\dagger} \phi^0$. The covariant derivative is

$$D_\mu \phi = \left(\partial_\mu + \frac{ig}{2} \vec{W}_\mu \cdot \vec{\tau} + \frac{ig'}{2} B_\mu \right) \phi. \qquad (4.42)$$

We again write $\phi^+ = (\phi_1 + i\phi_2)/\sqrt{2}$ and $\phi^0 = (\phi_3 + i\phi_4)/\sqrt{2}$ with the ϕ_i real. The potential is still given by (4.37) except that now the sums run from 1 to 4. For $\mu^2 < 0$ and $\lambda > 0$, the degenerate minima are on the surface of a four-dimensional sphere of radius $(\sum_{i=1}^4 v_i^2)^{1/2} = v = \sqrt{-\mu^2/\lambda}$. Without loss of generality,[21] we can choose $v_3 = v$ and $v_{1,2,4} = 0$. Substituting this in (4.41) leads to the mass terms

$$\mathcal{L}_\phi \to M_W^2 W^{+\mu} W_\mu^- + \frac{M_Z^2}{2} Z^\mu Z_\mu, \qquad (4.43)$$

---

[21] The labels ϕ^+ and ϕ^0 were chosen with this convention in mind, so that the background field would be electrically neutral and real.

where $W^{\pm} \equiv (W_1 \mp i W_2)/\sqrt{2}$ are the charged gauge bosons that mediate the WCC and Z is the predicted WNC neutral boson

$$Z \equiv \frac{-g' B + g W^0}{\sqrt{g^2 + g'^2}} = -\sin\theta_W B + \cos\theta_W W^0.$$

(4.44)

The *weak angle* θ_W is defined by $\tan\theta_W \equiv g'/g$, while the masses in (4.43) are

$$M_W = \frac{g v}{2}, \qquad M_Z = \frac{M_W}{\cos\theta_W}, \qquad (4.45)$$

implying the relation $\sin^2\theta_W = 1 - \frac{M_W^2}{M_Z^2}$. We will see later how g, v, and θ_W can be determined experimentally, leading to predictions for $M_{W,Z}$. For now, it suffices to state that the *electroweak scale* v is around $246\,\text{GeV} \sim 260\,m_p$, $M_W \sim 80.4\,\text{GeV}$, $M_Z \sim 91.2\,\text{GeV}$, and $\sin^2\theta_W \sim 0.23$. The combination of B and W^0 orthogonal to Z is the (massless) electromagnetic (photon) field

$$A = \cos\theta_W B + \sin\theta_W W^0. \qquad (4.46)$$

It can be shown that when quantizing around the vacuum the three Nambu-Goldstone bosons[22] associated with rolling on the surface of potential minima are eaten to become the longitudinal modes of the $W^{\pm}$ and Z.

---

[22]These are related to $\phi_{1,2,4}$ and analogous to θ in (4.38).

Therefore,

$$\phi \rightarrow \frac{1}{\sqrt{2}} \begin{pmatrix} 0 \\ v + H \end{pmatrix}, \qquad (4.47)$$

where H is the Higgs boson field describing radial vibrations. Using this full expression,

$$\mathcal{L}_\phi = \frac{1}{2} \left(\partial_\mu H \right)^2 - V(H) + M_W^2 W^{\mu+} W_\mu^- \left(1 + \frac{H}{v} \right)^2$$

$$+ \frac{1}{2} M_Z^2 Z^\mu Z_\mu \left(1 + \frac{H}{v} \right)^2. \qquad (4.48)$$

The first two terms are the kinetic energy and potential for H, where $V(H)$ is still given by (4.39) and has the same interpretation. The third and fourth terms are gauge boson mass terms and gauge-Higgs interactions. It is seen, for example, that there is a ZZH vertex proportional to M_Z^2/v. This proportionality to mass-squared (for bosons) or mass (for fermions) is characteristic of Higgs interactions.

The Fermions

We saw in section 4.5 that elementary chiral fermion mass terms such as $-m_d(\bar{d}_L d_R + h.c.)$[23] are not allowed by the (chiral) $SU(2) \times U(1)$ gauge symmetry because d_L and d_R transform differently. However, a Higgs-Yukawa

---

[23] $h.c.$ refers to the Hermitian conjugate $\bar{d}_R d_L$.

interaction

$$\mathcal{L}_d = -\sqrt{2}h_d\left(\bar{q}_L\phi\, d_R + h.c.\right) \qquad (4.49)$$

is consistent with gauge invariance. In (4.49), q_L is the $SU(2)$ doublet $\binom{u}{d}_L$, and the combination $\bar{q}_L\phi = \bar{u}_L\phi^+ + \bar{d}_L\phi^0$ of two $SU(2)$ doublets is $SU(2)$ invariant (as is d_R). The $U(1)$ charges also add up properly. h_d is the Higgs-Yukawa coupling constant, and we consider only one family. A similar term can be written relating $\ell_L \equiv \binom{\nu_e}{e^-}_L$ to e^-_R, while terms involving u_R and ν_{eR} require a slightly more complicated construction.

When ϕ is replaced by the SSB form in (4.47),

$$\mathcal{L}_d \to -m_d\left(\bar{d}_L d_R + \bar{d}_R d_L\right)\left(1 + \frac{H}{v}\right)$$

$$= -m_d\,\bar{d}d\left(1 + \frac{H}{v}\right), \qquad (4.50)$$

where $m_d = h_d v$. The d has acquired an effective mass due to its constant interaction with the background field, with the Higgs-Yukawa vertex proportional to m_d/v. Similar statements apply to the u, e^-, and ν_e.

The Higgs mechanism can be extended to three families. Then m_d becomes a 3×3 mass matrix,[24] which must be diagonalized just like a Hamiltonian in quantum

---

[24]The first term in (4.50) becomes $-\bar{d}_L m_d d_R(1 + H/v)$, where $d_{L,R}$ are three-component vectors consisting of the weak eigenstate fields, and m_d is a 3×3 matrix.

mechanics. The eigenvalues are the masses m_d, m_s, and m_b, and the corresponding fields $d_{L,R}$, $s_{L,R}$, and $b_{L,R}$ are known as *mass eigenstate* fields. They are unitary transformations of the *weak eigenstate* fields that we started with, analogous to energy eigenstates in quantum mechanics. (It will not be necessary for us to distinguish between them in our notation.) The couplings of the H to fermions are diagonalized by the same transformations, so

- The couplings are diagonal, i.e., they do not change one flavor into another. This is strongly supported experimentally, especially by the absence of large effects in the $K^0 - \bar{K}^0$ system. It is a strong constraint on theories with more complicated Higgs structures, which are not necessarily diagonal.
- The $\bar{f}fH$ couplings are $\propto m_f/v$: the Higgs-Yukawa coupling to the t is large; the couplings to b, c, and τ are considerably smaller; and the couplings to the first family and other light fermions are tiny. It is for this reason that the Higgs is difficult to produce and difficult to detect.

The Gauge Interactions

The fermion gauge interactions in (4.34) and the analogous lepton terms can be written in terms of the physical $W^\pm$, Z, and A after a bit of algebra as

$$\mathcal{L} = -\frac{g}{2\sqrt{2}} \left(J_W^\mu W_\mu^- + J_W^{\mu\dagger} W_\mu^+ \right)$$

$$- \frac{\sqrt{g^2 + g'^2}}{2} J_Z^\mu Z_\mu - \frac{gg'}{\sqrt{g^2 + g'^2}} J_Q^\mu A_\mu.$$

$$(4.51)$$

In the last term,

$$J_Q^\mu \equiv \frac{2}{3}\bar{u}\gamma^\mu u - \frac{1}{3}\bar{d}\gamma^\mu d - \bar{e}\gamma^\mu e, \qquad (4.52)$$

for one family. This is just the (parity-conserving) electro-magnetic current, so the term describes QED provided we identify

$$e = \frac{gg'}{\sqrt{g^2 + g'^2}} = g\sin\theta_W. \qquad (4.53)$$

The second term is the WNC interaction that was predicted by $SU(2) \times U(1)$, with

$$J_Z^\mu = \bar{u}_L\gamma^\mu u_L - \bar{d}_L\gamma^\mu d_L + \bar{\nu}_{eL}\gamma^\mu \nu_{eL}$$
$$-\bar{e}_L\gamma^\mu e_L - 2\sin^2\theta_W J_Q^\mu. \qquad (4.54)$$

It includes a parity-violating term, as well as a parity-conserving part proportional to $\sin^2\theta_W J_Q^\mu$, which led to the original experimental determination of $\sin^2\theta_W$. The $\sqrt{g^2 + g'^2}$ in the coefficient can be written as $g/\cos\theta_W = e/(\cos\theta_W\sin\theta_W)$.

The first term in (4.51) describes the WCC. The charge-raising and lowering currents include leptonic and hadronic contributions,

$$J_W^{\mu\dagger} = 2\bar{\nu}_{eL}\gamma^\mu e_L + 2\bar{u}_L\gamma^\mu d_L,$$
$$J_W^\mu = 2\bar{e}_L\gamma^\mu \nu_{eL} + 2\bar{d}_L\gamma^\mu u_L, \qquad (4.55)$$

which violate parity (and charge-conjugation) invariance. The vertices in (4.51) lead to typical electroweak processes shown in figure 3.2 on page 36.

The expressions for J_Q^μ and J_Z^μ can be extended to three families by simply summing over them. This extension is most elegantly denoted by redefining u to be a vector with components u, c, and t, and similarly for d, v, e, and their L - or R -chiral projections. For example, $\bar{u}\gamma^\mu u$ in J_Q^μ now represents

$$(\bar{u}\ \bar{c}\ \bar{t})\,\gamma^\mu \begin{pmatrix} u \\ c \\ t \end{pmatrix} = \bar{u}\gamma^\mu u + \bar{c}\gamma^\mu c + \bar{t}\gamma^\mu t. \quad (4.56)$$

J_Q^μ and J_Z^μ are diagonal in flavor, i.e., there are no terms like $-\bar{d}_L\gamma^\mu s_L$ in J_Z^μ. We will return to the historical and physical significance of this.

For J_Q^μ and J_Z^μ, we glossed over the distinction between the weak and mass eigenstate fermions: because of the diagonal nature of the currents, it didn't make any difference. For the WCC $J_W^{\mu\dagger}$, however, it does matter. The extension to three families retains the form of (4.55) for the weak eigenstate fields. However, the unitary transformations (mixings) between the weak and mass eigenstates are in general different for u_L and d_L, and also different for v_L and e_L, due to the mismatch between the gauge interactions and the Higgs-Yukawa interactions (or, equivalently, between the weak interactions and the physical masses relevant for QCD). The upshot is that the charge-raising current becomes

$$J_W^{\mu\dagger} = 2\bar{v}_L\gamma^\mu V_\ell e_L + 2\bar{u}_L\gamma^\mu V_q d_L, \quad (4.57)$$

where V_ℓ and V_q are 3×3 unitary matrices acting on the families. A similar form holds for J_W^μ except

$V \to V^{\dagger}$. The leptonic mixing matrix V_{ℓ} is not important for processes that are not sensitive to the (extremely tiny) neutrino masses, so we will ignore it for now. However, the quark mixing V_q is very important, and leads to family changing transitions such as $s \to u$, as well as family-conserving ones like $d \to u$. These had in fact been observed in strangeness-violating decays long before the SM was constructed.

V_q is known as the *Cabibbo-Kobayashi-Maskawa* (CKM) matrix. After removing the unobservable overall phases of the quark states, it depends on three angles and one phase. The latter is the only significant source of CP violation for quarks in the SM.[25] However, to a good first approximation the elements of V_q connecting the b and t to the first two families are small, so that

$$V_q \approx \begin{pmatrix} \cos \theta_c & \sin \theta_c & 0 \\ -\sin \theta_c & \cos \theta_c & 0 \\ 0 & 0 & 1 \end{pmatrix}. \qquad (4.58)$$

The upper 2×2 block in (4.58) is the *Cabibbo rotation*, which describes the relative strengths of the $d \leftrightarrow u$, $s \leftrightarrow u$, $d \leftrightarrow c$, and $s \leftrightarrow c$ transitions. It was known before the discovery of the third family that this form gives an excellent description of most WCC processes involving the first four quarks, with a *Cabibbo angle* of $\sin \theta_c \sim 0.23$. Nevertheless, the small corrections to (4.58) are needed to account for the observed CP violation and for b quark decays.

---

[25]The strong CP angle, to be discussed in chapter 5, is known to be small.

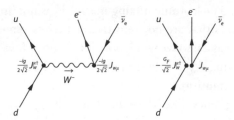

Figure 4.11. Left: Diagram for β decay at the quark level in the SM. Right: The zero-range amplitude in the Fermi theory, which is reproduced in the limit $|q^2| \ll M_W^2$. The two vertices are actually on top of each other, but are displaced for clarity. Source: Langacker 2010.

The Charged Current and the Fermi Theory

The quark-level amplitude for β decay is given by the first diagram in figure 4.11, with the vertices given by the charge-raising and lowering parts of $i\mathcal{L}$ in (4.51), and the W propagator by $-i/(q^2 - M_W^2)$, where q is the virtual four-momentum of the W. For β decay and most weak decays, $|q^2|$ is negligible compared to M_W^2, so that the propagator $\sim i/M_W^2$. This yields the same amplitude as one would obtain from the effective four-fermion interaction

$$\mathcal{L}_{Fermi} = -\frac{G_F}{\sqrt{2}} J_W^{\mu\dagger} J_W^{\mu}, \qquad (4.59)$$

where the Fermi constant G_F is given by

$$\frac{G_F}{\sqrt{2}} \equiv \frac{g^2}{8M_W^2} = \frac{1}{2v^2}, \qquad (4.60)$$

with the second form following from $M_W = g v/2$. $\mathcal{L}_{Fermi}$ is the zero-range effective interaction that Fermi wrote down in 1934 (updated to quark language and to include parity violation)! Although nonrenormalizable, the lowest-order Fermi interaction is an excellent first approximation to a large variety of WCC decays and scattering processes, so the $SU(2) \times U(1)$ model inherits all of these successes. Experimentally,[26] $G_F \sim 1.2 \times 10^{-5}$ GeV$^{-2} \sim 1.03 \times 10^{-5}$ m_p^{-2}. This is not enough information to determine g and M_W separately, but implies the value ~ 246 GeV for the electroweak scale $v = \sqrt{2}\langle 0|\phi^0|0\rangle$.

The processes successfully described by the Fermi theory include nuclear and neutron β decay, pion decay ($\pi^+ \to \mu^+ \nu_\mu$), muon decay ($\mu^- \to \nu_\mu e^- \bar{\nu}_e$), strange particle decays, heavy quark and lepton decays, and scattering processes such as $\nu_\mu e^- \to \mu^- \nu_e$. Observations of not only the rates but details such as the energy distributions and spin effects have confirmed the $V - A$ structure, to considerable precision in some cases. Comparison of different processes established the approximate form of the CKM matrix in (4.58) and measured the Cabibbo angle. Neutrino deep inelastic scattering ($\nu_\mu N \to \mu^- X$) verified the existence of spin-1/2 quarks, and a comparison with the DIS rate for $e^- N \to e^- X$ confirmed the electric charge assignments of the u and d quarks.

Let us elaborate a bit on the CKM matrix and its Cabibbo approximation. The underlying rates for β decay

---

[26]Even before the SM, there were many speculations on the existence of the $W^\pm$. The value of G_F suggests a mass of $\mathcal{O}(100$ GeV) if the coupling constant is anywhere close to e.

$(n \to p\, e^- \bar{\nu}_e)$ and muon decay $(\mu^- \to \nu_\mu\, e^- \bar{\nu}_e)$ were long known to be different by about 5% after correcting for kinematic and strong interaction effects, while strangeness-changing decays such as $\Sigma^- \to n\, e^- \bar{\nu}_e$ (Σ^- is an $\mathcal{S} = 1$ hyperon with mass $\sim 1.2\,\text{GeV}$) are much slower. This apparent nonuniversality in the strengths once appeared to be arbitrary, but became understandable in the context of the eightfold way (Cabibbo 1963) and even more so in the quark model. In that context, the leptonic and hadronic currents have the *same* overall strength, but the relevant charge $-\frac{1}{3}$ quark is not d but rather $d \cos\theta_c + s \sin\theta_c$. The rates for β decay ($d \to u\, e^- \bar{\nu}_e$), muon decay ($\mu^- \to \nu_\mu\, e^- \bar{\nu}_e$), and Σ^- decay ($s \to u\, e^- \bar{\nu}_e$) are therefore proportional to $\cos^2\theta_c \sim 0.95$, 1, and $\sin^2\theta_c \sim 0.05$, respectively.

More generally, *weak universality* is the principle that each WCC term has the same normalization in terms of weak eigenstates, and is only modified by fermion mixing effects. It is a necessary consequence of the $SU(2)$ gauge symmetry because every doublet has the same gauge interactions. One consequence is that the CKM matrix is unitary, $V_q V_q^\dagger = I$. The current value (Olive et al. 2014) of the 11 component is

$$\left(V_q V_q^\dagger \right)_{11} = |V_{ud}|^2 + |V_{us}|^2 + |V_{ub}|^2 = 0.9999(6),$$
$$(4.61)$$

in impressive agreement. The experimental values of the first two terms are obtained respectively by comparing superallowed β decay ($0^+ \to 0^+$ transitions between nuclei in an isomultiplet) and strangeness-changing decays such as $K^- \to \pi^0 e^- \bar{\nu}_e$ with muon decay.

The contribution of $|V_{ub}|^2 \sim 1.7(4)10^{-5}$, obtained from b decays, is negligible.

The agreement with universality is actually much more impressive than it first appears. That is because the the $SU(2) \times U(1)$ theory improves upon the Fermi theory by making it renormalizable, so that higher-order corrections (such as the exchange of a photon between the p and e^- in β decay) become calculable (Sirlin and Ferroglia 2013). The experimental precision underlying (4.61) is so high that these corrections are essential: the right-hand side would be ~ 1.04 if they were not included! In addition to testing the higher-order corrections, (4.61) also constrains some types of BSM physics, such as additional heavier gauge bosons with $V + A$ couplings or additional fermion families, which could modify the experimental quantities if they were not taken into account.

Not all processes involving the WCC can be approximated by the Fermi interaction. For example, physical (on-shell) $W^\pm$ can be produced in reactions such as $e^+e^- \rightarrow W^+W^-$ at very high energy, while $\Delta\mathcal{S} = \pm 2$ transitions like $K^0 \leftrightarrow \bar{K}^0$ can proceed via second-order weak processes involving two virtual W's. We will come back to both of these later.

The Weak Neutral Current

The WNC associated with the Z boson was the first major prediction of the $SU(2) \times U(1)$ model to be verified. It allowed for previously unobserved reactions such as elastic ($\nu_\mu e^- \rightarrow \nu_\mu e^-$ or $\nu_\mu p \rightarrow \nu_\mu p$) and deep inelastic

($\nu_\mu N \to \nu_\mu X$) neutrino scattering.[27] Such events were observed in the Gargamelle bubble chamber at CERN in 1973 and at Fermilab soon thereafter. Subsequent experiments studied WNC neutrino scattering in detail.

The WNC also makes new contributions to processes allowed by QED. At low energies, the WNC is tiny in comparison, but the axial vector couplings lead to effects that are absent or very small in QED. For example, the interference between Z exchange and the Coulomb interaction causes parity-violating mixing between atomic S and P wave states, leading to effects such as differences between the rates for transitions induced by left- and right-circularly polarized photons. These have been studied in detail in cesium and other atoms. There are also *polarization asymmetries*, i.e., relative differences in the cross sections for left- and right-helicity electrons scattering from electrons or nucleons, and angular asymmetries in $e^+ e^- \to \mu^+ \mu^-$. The WNC and its interference with the WCC contribution has also been measured in $\nu_e e^-$ and $\bar{\nu}_e e^-$ elastic scattering.

Generations of ever more precise WNC experiments, continuing to the present, have confirmed the predictions of the SM in great detail. They also allowed the value $\sin^2 \theta_W \sim 0.23$ to be determined utilizing the dependence of J_Z^μ on $\sin^2 \theta_W$ in (4.54) on page 97. One could then

---

[27]Most neutrino scattering experiments involve ν_μ or $\bar{\nu}_\mu$ rather than ν_e or $\bar{\nu}_e$ because the former are much easier to make. At accelerators, the neutrinos are mainly produced by charged pion decays like $\pi^+ \to \mu^+ \nu_\mu$, with the pions themselves emerging from protons interacting with nuclei in a target. The $V - A$ nature of the WCC (as opposed to scalar or tensor) implies that the $e^+ \nu_e$ rate is suppressed by a factor $(m_e / m_\mu)^2$, which is only partially compensated by phase space.

extract g from (4.53) and predict M_W and M_Z. The further implications of the WNC experiments will be discussed after we consider the W, the Z, and the Z-pole experiments.

The Z Pole

The measurement of $\sin^2 \theta_W \sim 0.23$ in the early WNC experiments, along with the known values of α and G_F, allowed the prediction of M_W and M_Z. Using the lowest-order results in (4.43), (4.53), and (4.60), one expects $M_W \sim 78\,\text{GeV}$ and $M_Z \sim 89\,\text{GeV}$, while higher-order corrections increase these predictions by $\sim 2\,\text{GeV}$. The W and Z were discovered by the UA1 and UA2 collaborations at the CERN SPS collider in 1983 in reactions such as $\bar{p}p \to W^+ + X$ with $W^+ \to e^+ \nu_e$, or $\bar{p}p \to Z + X$ with $Z \to e^+ e^-$. The masses and other properties were consistent with the SM expectations.

The standard model was tested to very high precision by experiments at the Large Electron-Positron Collider (LEP) and at the Stanford Linear Collider (SLC), which were $e^+ e^-$ colliders that operated at the Z-$pole$, i.e., the e^+ and e^- each had energy $\sim M_Z/2$ so that on-shell Z bosons could be produced and their decays studied. The cross sections at and near the Z-pole are extremely high, as can be seen for $e^+ e^- \to$ hadrons in figure 4.6 on page 77. LEP was a 27 km circumference circular $e^+ e^-$ collider at CERN that operated near the Z pole from 1989 to 1995, and at higher energy until 2000. The SLC, which ran from 1989 to 1998, collided e^+ and e^- beams that were

initially accelerated in the 2-mile SLAC linear accelerator
and then magnetically deflected to the collision point.
Four large detectors (ALEPH, DELPHI, L3, and OPAL)
surrounded the collision points at LEP, while a single
detector (initially MARK II and then SLD) observed the
collisions at the SLC. These and other detectors are enormously complicated beasts, which are highly instrumented
with devices that can track and characterize the particles
produced in a collision. The Z-pole experiments typically
involved hundreds of physicists from dozens of institutions
all over the world, and had dimensions of $\mathcal{O}(10 \text{ m})$ and
masses of $\mathcal{O}(1000 \text{ tons})$. The detectors at subsequent $\bar{p}p$
and pp colliders were even larger.

The LEP experiments together observed some 2×10^7
events near the Z-pole. The observables included the
Z mass and width, which can be determined from the
lineshape (cross section as a function of energy), as well
as the individual rates for $e^+e^- \to f\bar{f}$, where $f = e$,
μ, τ, b, or c, and the rate for $e^+e^- \to$ hadrons.[28] The
latter included the contributions of the u, d, and s quarks,
which could not be distinguished on an event-by-event
basis. In addition, LEP observed asymmetries sensitive to
the relative V and A couplings in (4.54). These included
forward-backward (FB) asymmetries, such as the fractional
difference between the number of μ^- moving in the same
and opposite hemispheres as the e^-, and the polarization

---

[28]The partial decay widths into each of these final states could be extracted
from these observations. Subtracting their sum from the total lineshape width
allowed an indirect determination of the rate for decays into neutrinos, which
could not be directly observed. The result was that there are only three light
ordinary neutrinos.

of $\tau^{\mp}$, determined from the angular distribution of their decay products. SLC had fewer events ($\sim 6 \times 10^5$) but had the advantage of a highly polarized e^- beam. This allowed measurement of the polarization asymmetries, which are especially sensitive to the value of $\sin^2 \theta_W$ and to perturbations from some types of BSM physics.

The Z mass was measured to be 91.1876(21) GeV at LEP, i.e., to the remarkable precision of $\sim 0.0023\%$, far more accurate than any other measurement at such high energies. This made use of a calibration of the beam energy by a resonant depolarization technique, and required corrections for the tidal effects of the Sun and Moon, the water table, the water level in Lake Geneva, and even for leakage currents from nearby trains! Most of the other measurements at LEP and the SLC were at the 0.1–1% level. They were generally in very good agreement with the SM predictions, although there were two discrepancies at the $\gtrsim 2\sigma$ level (Olive et al. 2014), possibly due to statistical fluctuations.

The full precision program included the WNC experiments; muon decay (needed to extract G_F); the Z-pole and higher-energy $e^+ e^-$ measurements[29]; precision measurements of lepton asymmetries at the Tevatron and LHC; direct measurements of M_W and m_t at LEP (in its later high-energy period), the Tevatron, and the LHC; and the measurement of M_H at the LHC. The precision of the measurements required an enormous theoretical effort to calculate higher-order corrections to all of the processes,

---

[29] LEP and the SLC also had extensive programs of QCD tests, heavy flavor physics, and searches for the Higgs boson and BSM physics.

such as those involving photon and gluon exchange. These included the running of α up to M_Z, which introduced the largest theoretical uncertainty because of hadronic effects. The results of the precision program included the following:

- The fermion interactions with the Z are consistent with the SM, supporting the gauge principle, the $SU(2) \times U(1)$ group, and the representations, i.e., that the left-chiral fermions are in $SU(2)$ doublets[30] and the right-chiral ones are singlets.

- Alternative electroweak theories (with different gauge groups) were excluded, and perturbations on the $SU(2) \times U(1)$ theory due, e.g., to additional particles or larger gauge groups, were strongly constrained. Many alternative models of spontaneous symmetry breaking involving new strong dynamics rather than an elementary Higgs boson predicted large (several %) deviations from the SM predictions and were excluded.

- The weak angle $\sin^2 \theta_W$ was determined to a precision of around 0.02%. The exact value depends on how the lowest-order definition $\theta_W = \tan^{-1}(g'/g)$ is generalized to higher orders.

- Higher-order corrections depend on parameters such as the masses of the top quark and the Higgs boson, which enter in vacuum polarization diagrams for the W and Z that are analogous to figure 4.7 and in the $Z \to b\bar{b}$ vertex. The corrections depend

---

[30]Asymmetries involving the τ lepton and the b quark implied the existence of the ν_τ and the t quark as their doublet partners, prior to their direct discovery.

quadratically on m_t/v and logarithmically on M_H/v, which allowed the t and H masses to be predicted prior to their direct discovery. Currently, these indirect predictions are $m_t = 177(2)$ GeV and $M_H = 89^{+22}_{-18}$ GeV, in reasonable agreement with the direct measurements $m_t = 173.2(0.9)$ GeV and $M_H = 125.7(0.4)$ GeV (Olive et al. 2014). Similarly, gluon exchange between the quarks in $Z \to q\bar{q}$ yields $\alpha_s(M_Z^2) = 0.1193(16)$, consistent with other determinations in figure 4.8.

- The success of the higher-order corrections vindicates the program of renormalization theory.
- The combination of α, $\sin^2\theta_W$, and α_s and the theoretical expressions for their running allow a test of whether the SM gauge interactions can be unified. That is, whether their running couplings meet at some high scale M_X, as would be expected if $SU(3) \times SU(2) \times U(1)$ is really part of a simpler group like $SU(5)$ that is spontaneously broken at scale M_X. As will be discussed in chapter 6, the observed gauge couplings are suggestive of a supersymmetric extension of the SM, with $M_X \sim 3 \times 10^{16}$ GeV.

Above the Z Pole

It was already mentioned in section 2.3 that the Fermi theory breaks down at high energies where cross sections grow so large as to violate unitarity. The same holds for the ad hoc extension to include massive $W^\pm$ bosons, known as the intermediate vector boson (IVB) theory. This problem

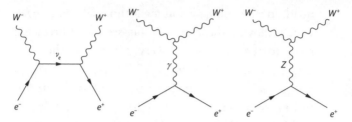

Figure 4.12. Diagrams for $e^+e^- \to W^+W^-$ in the SM, from Langacker 2010.

is cured in the SM (or other gauge theories) because the gauge invariance enforces delicate cancellations between Feynman diagrams that ensure renormalizability and an acceptable high-energy behavior. For example, in $SU(2) \times U(1)$ there are three lowest-order diagrams for $e^+e^- \to W^+W^-$, shown in figure 4.12. Only the first two, from ν_e exchange and annihilation into a photon, would be present in the IVB theory, leading to an unacceptable rapidly growing cross section. However, the third (Z annihilation) diagram in the SM is predicted to tame this behavior, in agreement with observations (Schael et al. 2013). The success of this prediction confirms the role of the Z and also that the ZW^+W^- and γW^+W^- vertices are consistent with the $SU(2)$ gauge self-interactions obtained from equation (4.28) on page 71. Other tests of the gauge self-interactions involve the pair production of electroweak gauge bosons at the Tevatron and LHC, and (for QCD) the running of $\alpha_s(Q^2)$.

Neutral Kaons, CKM, and *CP* Violation

The neutral kaons $K^0 = d\bar{s}$ and $\bar{K}^0 = \bar{d}s$ have definite strangeness $\mathcal{S} = +1$ and -1, respectively. They are

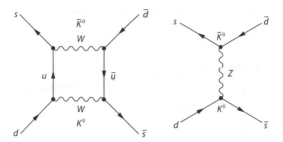

Figure 4.13. Left: A typical diagram leading to $\Delta S = \pm 2$ transitions between K^0 and $\bar{K}^0$ in the three quark (u, d, s) theory. In the SM with four or six quarks, there are additional diagrams in which each u is replaced by c or t. Right: An additional tree-level diagram in the three quark $SU(2) \times U(1)$ model. Source: Langacker 2010.

mapped into each other (up to a phase) by CP transformations and would be degenerate mass eigenstate particles in the absence of strangeness-violating weak interactions. However, the second-order weak interactions can cause transitions between the two, as in the first diagram in figure 4.13. This $K^0 - \bar{K}^0$ mixing has played two critical rules in the history of the weak interactions, which we consider in reverse chronological order.

The K^0 and $\bar{K}^0$ mass terms must be equal by the CPT theorem. Hence, any real (CP-conserving) $K^0 - \bar{K}^0$ transition amplitude implies that the actual mass eigenstates are the 45-degree admixtures

$$K_L = \frac{K^0 + \bar{K}^0}{\sqrt{2}}, \qquad K_S = \frac{K^0 - \bar{K}^0}{\sqrt{2}}. \quad (4.62)$$

Such mixing does occur. The mass difference $\Delta m_K \equiv m_{K_L} - m_{K_S}$ is measured indirectly (by effects somewhat similar to the neutrino oscillations to be described later) to be $\sim 3.5 \times 10^{-5}$ eV, which is incredibly small compared to the average mass of 498 MeV but still nonzero. However, there is a problem: a calculation of the first diagram in figure 4.13 (and ones related to it) in the W-boson (IVB) extension of the Fermi theory with the three quarks, (u, d, s) that were known prior to 1974 led to a value for Δm_K around 3000 times larger than the observed one! Matters became even worse in the $SU(2) \times U(1)$ model. With only three quarks it would be necessary to assign $d_L \cos \theta_c + s_L \sin \theta_c$ as the doublet partner of the u_L, while $-d_L \sin \theta_c + s_L \cos \theta_c$ and the right-chiral fields would be singlets. The second term in the WNC in (4.54) on page 97 would then contain a lowest-order *flavor changing neural current* (FCNC) piece proportional to $\bar{d}_L \gamma^\mu s_L + \bar{s}_L \gamma^\mu d_L$, leading to an even larger Δm_K through the second diagram in figure 4.13. It would also imply FCNC decays such as $K_L \to \mu^+ \mu^-$ at an unacceptable rate. This made it difficult to extend the original $SU(2) \times U(1)$ model of leptons (Weinberg 1967) to quarks.

The FCNC problem was resolved by the *Glashow-Iliopoulos-Maiani* (GIM) mechanism (Glashow et al. 1970), which postulated the existence of the charm quark. The c_L could then partner with the s_L (i.e., with $-d_L \sin \theta_c + s_L \cos \theta_c$) in a doublet so that J_Z^μ remained diagonal in flavor. The Z exchange diagram in figure 4.13 then disappears, while the higher-order diagrams with two W's are drastically reduced because of cancellations between the internal u and c quarks. The remainder is proportional to m_c^2 to an excellent approximation.

Comparison with the observed value led to the prediction (Gaillard and Lee 1974) $m_c \sim 1.5\,\text{GeV}$.

The GIM mechanism put the quarks and leptons on equal footing with four flavors each and elegantly resolved the FCNC problem. However, the quark model had been invented to simplify the hadronic spectrum, and many physicists were reluctant to complicate it with a new flavor. The situation changed dramatically when the J/ψ resonance was discovered at BNL and SLAC in late 1974 and was tentatively identified as a $c\bar{c}$ bound state, with $m_c \sim 1.5\,\text{GeV}$,[31] in agreement with the prediction from Δm_K from earlier that year! (See figure 4.6.) The identification was subsequently confirmed by the observation of mesons with a single c quark and by c production in neutrino scattering.

We now turn to CP violation (e.g., Kleinknecht 2003; Ibrahim and Nath 2008). K_L and K_S in (4.62) are eigenstates of CP with eigenvalues -1 and $+1$, respectively. It is straightforward to show that both $\pi^+\pi^-$ and $\pi^0\pi^0$ must have $CP = +1$ when their orbital angular momentum is zero. Therefore, if CP were exactly conserved, one could have $K_S \to 2\pi$, but $K_L \to 2\pi$ would be forbidden. K_L could still decay to 3π, but that should be much slower because of the smaller phase space. This is indeed observed: the K_S and K_L lifetimes are respectively $\sim 9 \times 10^{-11}\,\text{s}$ and $5 \times 10^{-8}\,\text{s}$, corresponding to decay lengths $c\tau$ of 3 cm and 15 m if they are relativistic. In fact, the subscripts S and L are for "short" and "long."

---

[31] This differs from the value 1.3 in table 3.1, because the latter is a running parameter evaluated at a different scale.

Following the discovery of space-reflection (P) and charge-conjugation (C) violation in the weak interactions in the 1950s, it was generally assumed that the product CP would still be a good symmetry of nature. It was therefore something of a shock when Val Fitch, James Cronin, and collaborators (Christenson et al. 1964) observed the decays of $K_L \rightarrow 2\pi$ in a neutral kaon beam at BNL, some 300 K_S decay lengths from the source, with an amplitude around 10^{-3} compared to that of $K_S \rightarrow 2\pi$. This suggested that $K_{L,S}$ are not the CP eigenstates given in (4.62), but instead have admixtures of the wrong CP state,[32] e.g.,

$$K_L = \frac{K^0 + \bar{K}^0}{\sqrt{2}} + \epsilon \frac{K^0 - \bar{K}^0}{\sqrt{2}}, \qquad (4.63)$$

where $|\epsilon| \sim 2 \times 10^{-3}$.

When the CP violation was first observed, there were speculations that its origin could be in some small contribution to the strong, weak, or (hadronic) electromagnetic interactions, none of which were well understood, or in an entirely new superweak interaction. However, it became even more puzzling after the development of the (two-family) SM because there seemed to be no room for it.

To understand this, we note that (4.62) follows automatically from CPT and the assumption that the

---

[32]This conclusion was correct, but not entirely justified at the time. The CP violation could have resided in the decay amplitude rather than in the states. Later experiments established that both are present, although the former is smaller by another factor $\sim 10^{-3}$.

$K^0 - \bar{K}^0$ transition amplitude is real.[33] A complex amplitude, however, will lead to the CP-violating form in (4.63), suggesting that CP violation is connected with complex phases in the interaction parameters. That indeed is the case, because pairs of CP-conjugate fields such as K^0 and $\bar{K}^0$ are basically Hermitian conjugates.[34] Since the Lagrangian must be Hermitian, interaction parameters involving K^0 or $\bar{K}^0$ must be complex conjugates of each other. Complex parameters could therefore lead not only to $\epsilon \neq 0$ but also to CP-violating differences between the K^0 and $\bar{K}^0$ properties.

Unfortunately, there are no observable phases in the SM for two families. The gauge coupling constants are real, and in an appropriate basis [such as $W^{\pm}$, W^0 for $SU(2)$ and the analog for $SU(3)$] the gauge fields couple to the currents with real coefficients. The 2×2 unitary matrices V_ℓ and V_q in (4.57) in principle involve three phases each. However, these are unobservable because they can be absorbed in the overall phases of the lepton and quark states. Finally, the Higgs-Yukawa couplings in (4.50) on page 95 are real in the mass eigenstate basis.

The absence of CP-violating phases was considered to be a serious difficulty in the early days of the SM. There remained the possibility of a superweak interaction or of a more complicated Higgs sector in which additional scalars have CP-violating interactions. However, the key came in the proposal of Makoto Kobayashi and Toshihide

---

[33] This discussion is somewhat oversimplified because it ignores the possibility of redefining the phases of the K^0 and $\bar{K}^0$. Observable CP-violating effects always involve interferences that are independent of such phase redefinitions.

[34] Additional Dirac matrices are involved for fermions.

Maskawa that there could be three families of fermions rather than two (Kobayashi and Maskawa 1973). The 3×3 unitary matrix V_q, now known as the CKM matrix, in general involves six phases, only five of which can be absorbed in the relative phases of the quark states. Thus, V_q involves three angles and one *observable CP*-violating phase, which originated in the Higgs-Yukawa couplings but migrated to V_q when going to the quark mass eigenstates.

The third family particles were eventually discovered, the τ lepton at SLAC in 1975, and the b and t at Fermilab in 1977 and 1995, respectively. The ν_τ was finally observed (by its rescattering to produce a τ) at Fermilab in 2000. The angles and CP violation associated with V_q have been studied extensively not only with the kaons, but also with c, b, and τ decays; neutrino scattering; and (to a lesser extent) t quark production and decays. There have also been major programs studying CP-violating mixings and decays in the analogs of the $K^0 - \bar{K}^0$ system involving heavier quarks, e.g., in $B_d^0 = d\bar{b}$, $B_s^0 = s\bar{b}$, and their CP-conjugates. These heavy quark studies have been carried out at many facilities, including the Cornell Electron Storage Ring (CESR), SLAC, the KEK laboratory near Tokyo, and at the Tevatron and LHC.

The upshot of all of these studies is that the three-family version of (4.57) successfully describes an enormous range of lowest and higher-order WCC processes. The elements connecting the first two families to the third turn out to be small, so (4.58) is a good first approximation for most CP-conserving effects involving the first two families, and the estimate of Δm_K is not too much affected by the t. However, the small residual elements lead to a phase in the $K^0 - \bar{K}^0$ transition and can account for the observed CP violation. The effect requires interferences between the box diagrams similar to figure 4.13 with u, c, and t exchanges.

One way to summarize the results is in the tests of the unitarity of V_q. The consistency of the weak universality prediction $\left(V_q V_q^\dagger\right)_{11} = 1$ with the data has already been displayed in (4.61). The *unitarity triangles* similarly test whether the off-diagonal elements of $V_q^\dagger V_q$ vanish. For example,

$$\left(V_q^\dagger V_q\right)_{31} = V_{ub}^* V_{ud} + V_{cb}^* V_{cd} + V_{tb}^* V_{td} \qquad (4.64)$$

is the sum of three complex numbers, which can be represented as two-dimensional vectors. Unitarity implies that their vector sum should vanish, i.e., that they should form a closed triangle. The lengths of the sides can be determined from CP-conserving effects such as the decay rates of mesons containing b and c quarks, or from the $B_d^0 - \bar{B}_d^0$ mixing, while the angles can be independently measured by various CP-violating asymmetries and decays. The overconstrained system is consistent with unitarity, as can be seen in figure 4.14. This not only confirms the CKM theory, but also constrains many types of new BSM physics that would lead to an apparent violation of unitarity if not included in the analysis. The agreement also provides an additional test of QCD because many strong interaction corrections and matrix elements for the various processes must be computed using lattice, perturbative, and other techniques.

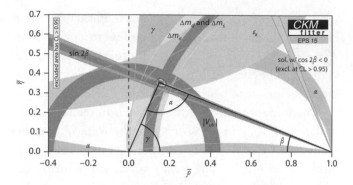

Figure 4.14. Constraints on the unitarity triangle prediction $\left(V_q^\dagger V_q\right)_{31} = 0$. The axes are the real and imaginary parts of the complex numbers in (4.64), scaled by the accurately known $|V_{cb}^* V_{cd}|$. The various constraints are consistent with the closing of the triangle. Plot courtesy of the CKMfitter group (Charles et al. 2005, hep-ph/0406184). Updated results and plots available at http://ckmfitter.in2p3.fr.

The relatively large angles in figure 4.14 imply that in some sense, CP violation is large. The small value of ϵ in (4.63) and of other CP-violating effects is not due to a small phase, but because of the small mixing angles that connect the t to the d and s. With the enormous benefit of hindsight, CP violation is perhaps not so surprising as it seemed in 1964. After all, quantum mechanics is based on complex numbers, which can lead to CP breaking. The real issue is whether a given system or process is sufficiently complicated for the phases to lead to observable interference effects. In QCD,[35] QED, the WNC, and

---

[35]The strong CP effect to be discussed in chapter 5 is of a completely different character.

two-family WCC, there are no such phases. Only the three-family theory involves an unremovable phase, and even there CP violation requires that all three families are relevant to the process. The possibility of an analog to the CKM phase in the leptonic mixing will be mentioned later.

The Large Hadron Collider (LHC)

By the time the LHC started running, most of the major ingredients of the standard model had been verified, including the fermion gauge interactions, the W and Z, the unitarity of the CKM matrix, and even the higher-order corrections. The outstanding ingredient was the mechanism of electroweak symmetry breaking, i.e., is it due to the minimal (one-doublet) Higgs mechanism, a more complicated Higgs sector, or some dynamical or compositeness mechanism not involving elementary scalars?

The LHC is a pp collider at CERN with a design CM energy of 14 TeV installed in the same 27 km circumference deep underground tunnel that had previously housed LEP. Its primary goals were to search for the Higgs boson and to search for the possible BSM physics at the TeV scale that was motivated by the unanswered questions to be considered in chapter 5.

The LHC startup was delayed by more than a year by an unfortunate accident caused by a faulty electrical connection that damaged many of the superconducting magnets. It restarted in 2010 and ran at 7 and then 8 TeV

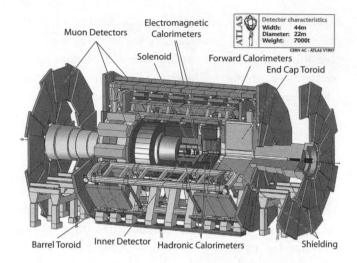

Figure 4.15. A schematic view of the 44 m wide, 22 m diameter, and 7000 ton ATLAS detector, courtesy of CERN. The height of a typical human is about the same size as the capital letters of the words identifying the various parts of this mechanism.

CM energy through early 2013. Following a shutdown for upgrades, the LHC began a new higher-energy run at $\sim 13\,\text{TeV}$ in the spring of 2015.

There are four large detectors located at intersection points of the beams and several smaller experiments. ATLAS and the Compact Muon Solenoid (CMS), shown respectively in figures 4.15 and 4.16, are huge general-purpose detectors, each built and operated by thousands of physicists and engineers from nearly 200 institutions in dozens of countries. They are highly instrumented to identify and track the large numbers of particles produced in the pp collisions, and involve sophisticated electronic

Figure 4.16. A view of part of the CMS detector, opened so that the inside is visible. Peter Higgs is in the foreground. Figure courtesy of CERN.

readout systems and worldwide distributed computing networks for the data analysis. The LHCb detector is optimized for b-quark physics, while ALICE is designed to study the plasma of free quarks and gluons that emerge at high temperatures and densities in heavy ion collisions.

The Higgs Boson

From expression (4.39) on page 89 and $v \equiv \sqrt{-\mu^2/\lambda}$, one finds $M_H^2 = 2\lambda v^2$ for the square of the Higgs mass. $v \sim 246\,\text{GeV}$ is known through the relation to the Fermi constant in (4.60). However, there is no a priori knowledge of λ, at least in the lowest order, except that it is nonnegative from vacuum stability. M_H is apparently completely arbitrary.

The situation is improved somewhat when one takes into account higher-order corrections, which lead to a running $\lambda(Q^2)$, with $M_H^2 = 2\lambda(v^2)\,v^2$. For large M_H, the running is dominated by diagrams involving intermediate Higgs fields, leading to an *increase* of $\lambda(Q^2)$ with Q^2. Eventually, $\lambda(Q^2)$ diverges. This would not matter if the SM is just an approximation to a more fundamental theory that takes over at some new physics scale Λ_{NP} and if $\lambda(Q^2)$ remains finite for $|Q| < \Lambda_{NP}$. However, it would be a disaster[36] if the divergence scale is within the domain of validity of the SM, i.e., if it is smaller than Λ_{NP}.

For smaller values of M_H, the running is instead dominated by diagrams involving virtual top quarks, because of the rather large Higgs-Yukawa coupling m_t/v to the top. In this case, $\lambda(Q^2)$ *decreases* with Q^2 and eventually goes negative. The corrections make the vacuum unstable or possibly metastable if this zero point is smaller than Λ_{NP}. Metastability would be acceptable (though possibly uncomfortable!) if the lifetime is longer than the known 14 billion year age of our Universe.

For the SM to be valid up to to the Planck scale, $\Lambda_{NP} = M_P$, the twin requirements of finiteness and vacuum stability restrict the Higgs mass to the approximate range 130–180 GeV, with the lower limit weakened to ~ 115 GeV if the vacuum is allowed to be metastable but long-lived. If the SM breaks down at $\Lambda_{NP} \sim 1.5$ TeV, on the other hand, the much larger range 85–650 GeV would be possible.

---

[36]Or at least it would be inconsistent with the implicit assumption that the SM fields can be treated as elementary and weakly coupled below Λ_{NP}.

It is interesting that in the supersymmetric extension of the SM λ is no longer a free parameter but is instead related to the $SU(2) \times U(1)$ gauge couplings. In the simplest supersymmetric extension with TeV-scale masses (so that $\Lambda_{NP} \sim$ TeV), there is an additional *upper* bound of ~ 130 on the lightest Higgs scalar, which can be relaxed somewhat in more complicated versions. Prior to the Higgs discovery, it had been hoped that an observation of $M_H \gg 130$ GeV would point toward a large new physics scale, while $M_H \ll 130$ GeV would suggest supersymmetry. The value 125 GeV that was ultimately found is inconclusive and somewhat challenging for both views. This is symptomatic of what is perhaps the key question in particle physics: is the new physics scale in the TeV range, or does the SM work up to much higher energies?

The interactions of the Higgs to other SM particles are proportional to mass (for fermions) or mass-squared (bosons), as in (4.50) and (4.48). The largest couplings are therefore $\bar{t}tH$, ZZH, and W^+W^-H, and the dominant production mechanisms in both e^+e^- and hadron colliders involve these vertices. Similarly, the H decays predominantly into the heaviest particles that are kinematically possible. If M_H had been larger than $2M_W$, for example, then $H \to W^+W^-$ would have dominated, while for the actual mass of 125 GeV the largest branching ratio is into $b\bar{b}$. However, the $b\bar{b}$ decay is difficult to observe in pp or $\bar{p}p$ collisions because of the much larger background rate of $b\bar{b}$ production by QCD processes, so the actual Higgs discovery relied on much rarer but cleaner decay channels.

It has already been mentioned on page 109 that the precision electroweak program led to a prediction of

M_H through the higher-order corrections to the W and Z propagators. The current value is $M_H = 89^{+22}_{-18}$ GeV, somewhat lower than 125 GeV but consistent at 1.6σ. Furthermore, direct searches for the Higgs were carried out at LEP, which eventually achieved a lower limit $M_H > 114.4$ GeV through the nonobservation of $e^+e^- \to Z^* \to ZH$, where Z^* is virtual.

The Higgs was subsequently sought by the CDF and D0 collaborations at the Tevatron ($\bar{p}p$) and later by ATLAS and CMS at the higher-energy LHC (pp). The dominant production mechanism was *gluon fusion*, in which $GG \to t\bar{t}$ followed by $t\bar{t} \to H$. *Higgstrahlung* ($V^* \to VH$, where $V = W$ or Z) and VV *fusion* ($VV \to H$, with the V radiated from quarks) also contributed. CDF and D0 were initially able to exclude masses in the 160–170 GeV region, and later the ATLAS and CMS had excluded the whole range from 127 to 600 GeV, as well as below 116. This left only a fairly narrow window, and many physicists became quite skeptical that the SM Higgs existed. However, that window was on the upper edge of the region suggested by precision electroweak, and was also consistent with the supersymmetry expectation.

By the end of 2011, however, both LHC experiments observed significant excesses around 126 GeV in the $\gamma\gamma$ and 4ℓ channels, where $\ell = e^\pm$ or $\mu^\pm$. (A candidate 4μ event is shown in figure 4.17.) These were suggestive of $H \to \gamma\gamma$, which proceeds via an intermediate loop of t or W, and $H \to ZZ^* \to 4\ell$. Both of these channels have small branching ratios, because of the need for higher order for $\gamma\gamma$ or because of the virtual Z^* for 4ℓ, but both have clean signatures and could be separated from

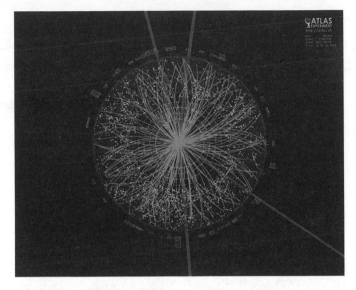

Figure 4.17. A candidate $H \to ZZ^* \to 4\mu$ event from the ATLAS experiment. The muon tracks are those exiting the central region. Figure courtesy of CERN.

the large backgrounds from other processes. None of the observations yet had the 5σ significance required to claim a discovery in particle physics, but they were highly suggestive.

Considerably more data and analysis effort had been accumulated six months later. At 9:00 AM European summer time on 4 July 2012, a presentation of the results was given in the CERN auditorium and broadcast worldwide over the Internet (See The photograph in figure 4.18). The timing was chosen to coincide with the International Conference on High Energy Physics that was

Figure 4.18. At CERN after the announcement of the Higgs-like boson on 4 July 2012, courtesy of CERN. From left: Fabiola Gianotti (ATLAS), Rolf Heuer (Director General), Joe Incandela (CMS).

taking place in Melbourne, Australia, but was somewhat inconvenient for those of us in the Americas. Nevertheless, the seminar room at the Institute for Advanced Study was packed at 3:00 AM on the US Independence Day to watch the anticipated announcement. I was personally rather sleepy at the beginning, but quickly became wide awake as first Joe Incandela, leader of the CMS effort, and then Fabiola Gianotti, the ATLAS head, presented their results. Each group had found significant signals in $\gamma\gamma$ and 4ℓ with mass around 125–126 GeV. The individual measurements were each a bit short of 5σ, but the collection of results left little doubt. Rolf Heuer, the Director General of CERN, summed it up nicely: "As a layman, I would now say I think we have it. Do you agree?"

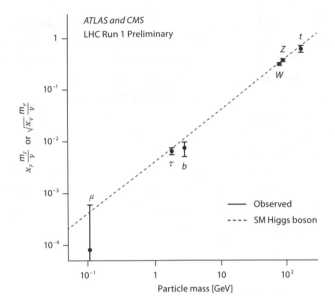

Figure 4.19. Observed couplings of the Higgs boson, as a function of the particle mass. The data are consistent with the expectation that the points should fall on a straight line. Courtesy of CERN.

At the IAS, we followed by a reception, with champagne thoughtfully provided by Nima Arkani-Hamed.

Additional running and analysis of LHC data[37] allowed more detailed studies of the production mechanisms, additional decay channels, and angular distributions; increased the significance well above 5σ; and determined the mass to be 125.09 ± 0.24 GeV (Aad et al. 2015; Murray and Sharma 2015). The observed particle has spin-0 and even parity, with couplings consistent with the Higgs expectations (figure 4.19). It is clearly either the minimal SM

---

[37] And of an excess in the Tevatron data.

Higgs or something close to it. Nevertheless, the statistical precision of $\sim 20\%$ still leaves room for deviations from the SM, e.g., because of an extended scalar sector or because the Higgs is actually composite. Both of these possibilities will be mentioned in chapter 6.

As mentioned earlier, the ~ 125 GeV mass is marginally consistent with supersymmetry, though somewhat high. Conversely, it is somewhat low if there is no new physics up to a high scale: it sits in the narrow metastability band between vacuum stability and instability,[38] with $\lambda(Q^2)$ going negative around $10^{10} - 10^{12}$ GeV. Perhaps nature is trying to tell us something.

Neutrino Mass and Mixing

The neutrinos are the oddballs of particle physics.[39] They were long believed to be massless, and the original $SU(2) \times U(1)$ model treated them as such. In fact, the right-chiral gauge singlet fields ν_R in table 3.1 were not even included. However, we now know from *neutrino oscillations* that at least two of the neutrinos have nonzero mass, but these are very much smaller than even that of the electron. It is straightforward to extend the SM to allow for neutrino mass, but there are actually two different ways to do so: *Dirac masses*, analogous to the quark and charged lepton masses, or *Majorana masses*, which violate lepton number by two units. It is perhaps surprising that the tiny

[38] Fortunately, most estimates are that the decay rate is comfortably low!
[39] For a general review, see Barger et al. 2012.

neutrino masses may be associated with BSM physics, such as grand unification or superstring theory, at a very large scale.

All weak decay[40] and scattering experiments were consistent with massless neutrinos, and in fact no existing experiment is sensitive to the kinematic effects of masses in the sub-eV scale. However, there is much more sensitivity in neutrino oscillations, in which even tiny mass differences can lead to transformations of one type of neutrino into another. Neutrino oscillations are due to the mismatch between weak and mass eigenstates. They are an example of the oscillations that occur in any quantum system in which the initial state is a superposition of energy eigenstates, such as in the vibrations of an ammonia molecule, or in a classical system involving weakly coupled harmonic oscillators. Suppose, for example, that the flavor eigenstate fields ν_e and ν_μ, associated with the e^- and μ^- respectively in weak transitions, are linear combinations of the mass eigenstates $\nu_{1,2}$, as in equation (4.57) on page 98 with $V_\ell \neq I$, so

$$\begin{aligned} |\nu_e\rangle &= |\nu_1\rangle \cos\theta + |\nu_2\rangle \sin\theta, \\ |\nu_\mu\rangle &= -|\nu_1\rangle \sin\theta + |\nu_2\rangle \cos\theta \end{aligned} \qquad (4.65)$$

for the corresponding states. Suppose further that at time $t = 0$ the state $|\nu(0)\rangle = |\nu_\mu\rangle$ is created by the

---

[40]With the exception of some suggestions from β decay that turned out to be erroneous.

decay $\pi^+ \to \mu^+ \nu_\mu$. At a later time $t > 0$, this will have evolved into

$$|\nu(t)\rangle = -|\nu_1\rangle \sin\theta\, e^{-iE_1 t} + |\nu_2\rangle \cos\theta\, e^{-iE_2 t},$$
(4.66)

where $E_{1,2}$ are the energies of the two mass eigenstates. $|\nu(t)\rangle$ is a superposition of $|\nu_e\rangle$ and $|\nu_\mu\rangle$. If one places a target a distance $\mathcal{L}$ from the source, the $|\nu_e\rangle$ component may rescatter to produce an e^-, i.e., it may have oscillated into ν_e with probability

$$P_{\nu_\mu \to \nu_e} = |\langle \nu_e | \nu(t) \rangle|^2 = \sin^2 2\theta \sin^2\left(\frac{\Delta m^2 \mathcal{L}}{4E}\right),$$
(4.67)

where $\Delta m^2 \equiv m_2^2 - m_1^2$. We have assumed the neutrino is extremely relativistic,[41] with average energy $E \gg m_{1,2}$ and $\mathcal{L} \sim t$. The oscillation formula is independent of whether the masses are Dirac or Majorana, and can easily be extended to three flavors.

The first hint of neutrino oscillations came from Ray Davis' ^{37}Cl experiment, designed to detect ν_e's produced in nuclear reaction chains in the core of the Sun.[42] The experiment involved 10^5 tons of cleaning fluid in a tank deep underground (to shield from cosmic rays) in the Homestake gold mine in South Dakota. The Solar neutrinos were detected by the reaction $\nu_e + {}^{37}Cl \to e^- + {}^{37}Ar$, with the argon atoms detected by their decays

---

[41] The energies are $E_i = (|\vec{p}|^2 + m_i^2)^{1/2} \sim |\vec{p}| + m_i^2/2|\vec{p}|$, where we have assumed a common momentum $|\vec{p}| \sim E$.

[42] I first learned about this experiment in a lecture by Philip Morrison at MIT around 1965, the first physics colloquium that I ever attended.

after being separated chemically. By the early 1970s, it became apparent that Davis was observing only about 1/3 of the v_e flux that was predicted by the *standard Solar model* (SSM) calculations of John Bahcall and others (Bahcall 1989). One possibility was that some of the v_e's were transforming into other flavors to which the experiment was not sensitive. The other was that the SSM was at fault. For example, the experiment was mainly sensitive to the highest energy v_e's expected from the Sun, especially from the relatively rare process $^8B \rightarrow {}^8Be^* + e^+ + v_e$, and their predicted flux was extremely sensitive to the temperature of the Solar core.

It took some 30 years to definitely resolve the situation, which required the simultaneous sorting out of the Solar and the neutrino physics. This required a number of different Solar neutrino experiments, involving interactions in ordinary and heavy water, gallium, and liquid scintillators. These were sensitive to different parts of the Solar spectrum and therefore to different parts of the reaction chain, and the heavy water experiment (the Sudbury Neutrino Observatory [SNO] in Canada) even allowed a measurement of the sum of all three flavors through neutral current scattering. There were also many refinements to the theory, measurements of relevant nuclear cross sections, and independent probes of the Solar interior through helioseismology (the study of acoustic waves propagating through the Sun). In the end, the SSM was vindicated; v_e really are transforming[43] into v_μ and

---

[43] It turns out that the transformations of the higher-energy Solar neutrinos are governed not by the *vacuum oscillations* in (4.67) but by an interplay between

ν_τ before reaching the Earth. The Solar neutrinos indicate a tiny splitting, $\Delta m_\odot^2 = m_2^2 - m_1^2 \sim 7.5 \times 10^{-5}$ eV^2 between two mass eigenstates, with a mixing $\sin\theta_\odot \sim 0.55$ that is large compared to the quark sector Cabibbo angle ($\sin\theta_c \sim 0.23$).

The Solar neutrinos were the first serious evidence for neutrino mass. However, the first definitive discovery of neutrino oscillations was from the *atmospheric neutrinos*, which are produced by the decays of pions and other particles produced by cosmic ray collisions in the atmosphere. The *SuperKamiokande* experiment, located under a mountain in Western Japan (to shield from cosmic rays), consists of 50 kilotons of ultrapure water surrounded by photomultiplier tubes to detect Cerenkov radiation from the products of neutrino interactions and proton decay. SuperKamiokande and its predecessor Kamiokande found anomalies in the ratio of ν_μ to ν_e atmospheric neutrinos, and especially in the dependence of the fluxes on the distance they had traveled (which ranged from ~ 15 km, for cosmic ray interactions overhead, to the diameter of the Earth). By 1998, they had accumulated sufficient statistics to conclusively establish neutrino oscillations (and thus quantum-mechanical interference effects on a distance scale of $\mathcal{O}(10^4$ km)!). These are mainly between ν_μ and ν_τ, with $|\Delta m_{atm}^2| = |m_3^2 - m_2^2| \sim 2.5 \times 10^{-3}$ eV^2. The mixing is consistent with maximal ($\pi/4$), i.e., $\sin\theta_{atm} \sim 1/\sqrt{2}$.

---

the neutrino mass effects and coherent forward scattering from matter in the Sun. Unlike vacuum oscillations, these matter effects allow determination of the sign of Δm^2.

The Solar and atmospheric neutrino results have been confirmed by a number of terrestrial experiments with neutrino beams produced in accelerators and reactors. For example, the Solar results were confirmed by the Kamland experiment, which involved the observation of $\bar{\nu}_e$'s from a number of Japanese reactors located around 200 km from a liquid scintillator detector, while the atmospheric and other results have been studied in long baseline (hundreds of km) experiments in Japan, Europe, and the United States. We have seen that, unlike the quarks, two of the leptonic mixing angles are large. For a time, it appeared that the third angle might be zero, and a number of theoretical models attempted to motivate this. Eventually, however, several experiments, most precisely the Daya Bay and RENO reactor experiments in China and South Korea, respectively, have established a small but nonzero value, $\sin \theta_{13} \sim 0.15$.

The oscillation results imply that at least two of the three neutrinos have nonzero mass. Since atmospheric oscillations determine only the magnitude of Δm^2_{atm}, it it is not known whether m_3 is larger or smaller than m_2 and m_1. These cases are referred to as the *normal* and *inverted hierarchy*, respectively. If, for example, the smallest mass were zero, then one would have $m_1 = 0$, $m_2 \sim 0.009$ eV, and $m_3 \sim 0.050$ eV for the normal hierarchy, while $m_3 = 0$, $m_1 \sim 0.049$ eV, and $m_2 \sim 0.050$ eV for the inverted.

There is no reason to expect the lightest neutrino to be exactly massless, but the overall scale cannot be too large. Oscillations depend only on the differences in m^2_i, but other effects depend on the masses themselves.

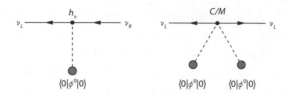

Figure 4.20. Left: Generation of a Dirac mass by the Higgs mechanism. Right: Generation of a Majorana mass by a higher-dimensional operator involving two Higgs fields. Source: Langacker 2010.

Early limits were obtained from kinematic effects in β decay, π decay, and τ decay. Currently, however, the most stringent constraints are from cosmology. Light neutrinos could have freely streamed away from density perturbations in the early Universe, modifying the CMB (see section 3.3) and galaxy distributions. The Planck collaboration obtains an upper limit of 0.23 eV on the *sum* of the neutrino masses (Ade et al. 2015), not all that much larger than the scale suggested by neutrino oscillations.

There is still much that is not understood about the neutrino masses, especially whether they are Dirac or Majorana. A Dirac mass could be generated in the same way as those of the quarks and charged leptons, by coupling the left-chiral lepton doublet ℓ_L to the right-chiral singlet neutrino ν_R and the Higgs doublet, similar to (4.50) on page 95 and illustrated in figure 4.20. The neutrino part would be

$$\mathcal{L}_\nu = -\sqrt{2}h_\nu \left(\bar{\nu}_L\phi^{0*}\nu_R + h.c.\right) \rightarrow -m_\nu(\bar{\nu}_L\nu_R + h.c.),$$
$$(4.68)$$

with $m_\nu = h_\nu v$. The problem is that the needed Higgs-Yukawa coupling would have to be extremely small,

$h_\nu \sim 10^{-12}$ for $m_\nu \sim 0.1$ eV, and many people feel that there should be some explanation for such a small parameter.[44]

Small Majorana masses could have a simpler explanation. A Majorana mass term for a left-chiral neutrino is of the form $\frac{1}{2} m_\nu \, \nu_L \nu_L$ (or, more precisely, $\frac{1}{2} m_\nu \, \ell_L \ell_L$), up to some Dirac and $SU(2)$ matrices. A $\nu_L \nu_L$ term can be thought of as turning a neutrino into an antineutrino, or equivalently as creating or annihilating two neutrinos. It therefore violates fermion and lepton numbers by two units.[45] It also violates $SU(2)$ by a full unit. Assuming no new Higgs fields, it would have to emerge from a *higher-dimensional operator* of the form (Weinberg 1980b)

$$\mathcal{L}_\nu = \frac{C}{M}(\ell_L \ell_L \phi \phi + h.c.) \rightarrow \frac{C v^2}{2M}(\nu_L \nu_L + h.c.),$$
$$(4.69)$$

where C is a dimensionless coefficient and M is a new mass scale, presumably associated with BSM physics. Such an operator is not renormalizable, but may emerge as a low-energy approximation to some more fundamental theory, much like the Fermi interaction in (4.59) is an approximation to the WCC in $SU(2) \times U(1)$. In this case, $m_\nu \sim C v^2/M$, and a tiny m_ν would ensue for $M/C \gg v$, e.g., $M/C \sim 10^{14}$ GeV. C is unknown, but is often taken to be $\mathcal{O}(1)$.

---

[44]Of course, the larger but still tiny value, $h_e \sim 10^{-5}$, of the Higgs-Yukawa coupling for the electron is not understood either, nor is the hierarchy $h_e \ll h_t \sim 0.7$.

[45]Analogous terms for the charged leptons and quarks are not allowed because they would violate the conservation of electric charge (and also of color for the quarks).

The simplest implementation of these ideas is the *seesaw model,* in which the right-chiral ν_R obtains a large Majorana mass of $\mathcal{O}(M)$ from new physics, while an ordinary-sized Dirac mass $m_D \sim h_\nu \nu$ (comparable to a quark or charged lepton mass) connects ν_L and ν_R. This results in a 2×2 mass matrix, with approximate eigenvalues M and $m_D^2/M = h_\nu^2 \nu^2/M \ll m_D$, i.e., the light neutrino is naturally very light for large M.

Other theoretical ideas for generating small Dirac or Majorana masses, the possibility of distinguishing between them by searching for the *neutrinoless double beta decay* $(\beta\beta_{0\nu})$ process $nn \rightarrow ppe^-e^-$, and the possible connection of the seesaw model with baryogenesis will be briefly considered in chapter 6.

Other open questions for the neutrinos include the absolute mass scale, the hierarchy, and the possibility of leptonic CP violation. Just as for the quarks, CP breaking could result from the observable phase in the 3×3 leptonic mixing matrix V_ℓ in (4.57) on page 98, which is known as the *Pontecorvo, Maki, Nakagawa, Sakata* (PMNS) matrix. Neither the magnitude of the phase nor even whether it is nonzero are known at present. Finally, there are some intriguing hints that there may be additional *sterile* $[SU(2)\text{-singlet}]$ neutrinos in the eV range that mix with the ordinary neutrinos, but there are also strong constraints on at least the simplest sterile neutrino scenarios from other experiments and cosmology.

5

WHAT DON'T WE KNOW?

A story went around when I was a graduate student at
Berkeley: at a lunch seminar a few years earlier a well-
known physicist had remarked that he was determined
to understand the strong interactions during his lifetime,
whereupon one of his colleagues quipped that that was a
problem for medicine and not for physics. I don't know
whether the incident really occurred (I suspect that it
did), but it certainly reflects the challenges faced at the
time in understanding the strong (and weak) interactions.
Fortunately, however, by some combination of inspiration,
perspiration, and the cooperation of nature, we have now
developed and tested the standard model (SM), which
really does describe almost everything we observe to an
excellent approximation. Nevertheless, I don't know of
anyone who believes that the SM is the final story. It is
just too complicated in detail and has too many aspects
that must be taken from experiment rather than explained
from first principles.

In this chapter, I will examine these issues, first describing features that appear to be arbitrary or fine-tuned,

then discussing some of the questions that are not really addressed by the SM, and finally examining some of the paradigms (such as uniqueness and minimality) that are generally assumed but are starting to be questioned.

5.1 Arbitrariness and Tuning

The standard model consists of the $SU(3) \times SU(2) \times U(1)$ gauge theory for the microscopic interactions and classical general relativity for gravity. As described in the previous chapter, most aspects have been experimentally verified, often to high precision. However, even though the basic idea of a gauge theory is simple, the actual implementation in the SM is extremely complicated. One aspect is that the current version of the SM (three families with massive neutrinos) involves 27 or 29 (depending on whether the neutrino masses are Dirac or Majorana) fundamental parameters[1] that must be taken from experiment. One might hope that an ultimate theory would somehow explain the values of these parameters, e.g., from some geometric construction or as solutions to algebraic equations.

The complications are also summarized under what I call the "five problems" (Langacker 2010).

---

[1] The fermion masses, mixing angles, and CP phases; the gauge couplings; the electroweak scale v; the Higgs mass; the strong CP angle (introduced later); the Planck scale; and the cosmological constant; minus one, since only mass ratios are observable. The SM does not include the dark matter.

The Interactions Are Complicated

The strong, weak, and electromagnetic interactions are all described by gauge theories, but in fact they are very complicated. There are three commuting gauge factors, $SU(3)$, $SU(2)$, and $U(1)$, each with its own gauge-coupling constant. The $SU(3)$ symmetry and one linear combination of the $SU(2) \times U(1)$ symmetries are unbroken, while the other $SU(2) \times U(1)$ symmetries are broken by the Higgs mechanism. Similarly, the unbroken symmetries are parity conserving, while the others are parity violating.

Furthermore, there is no fundamental explanation of *charge quantization*,[2] i.e., that all of the fundamental particles have electric charges that are multiples of $e/3$. The fermion electric charges are related to the underlying $U(1)$ charge assignments by equation (4.32) on page 85, but the latter were actually chosen to give the observed values and not predicted. The charge assignments could arguably have been included in the counting of free parameters, but I did not do so because of their observed discrete values and because of several constraints from the cancellation of quantum pathologies known as *anomalies*.[3]

Charge quantization suggests some form of unification. In grand unification, for example, the SM is embedded in a larger gauge group, such as $SU(5)$, that does not have commuting factors. Superstring theory, which incorporates quantum gravity, also leads to charge quantization,

---

[2] An important consequence of charge quantization is the electrical neutrality of atoms, i.e., that the electron and proton charges are equal and opposite and the neutron is neutral. This has been verified experimentally at the $10^{-21}e$ level.

[3] Anomaly cancellation is not sufficient to determine all of the charges without additional assumptions.

though not all string vacua correspond to the observed charges.

The Spectrum Is Complicated

Even though the gauge interactions are complicated, they are relatively constrained compared to the Higgs-Yukawa couplings. The latter are completely arbitrary at our present level of understanding, with the consequence that the masses, mixings, and CP violation of the quarks and leptons are simply not understood. Even the number of fundamental fermions is a mystery. These issues are known collectively as the *flavor problem*.

The u and d quarks and the electron are the only fermions that are needed for ordinary matter under normal circumstances. The ν_e is arguably essential for nucleosynthesis, and antiparticles are a necessary consequence of quantum mechanics and relativity. However, the heavier families, $(c, s; \nu_\mu, \mu^-)$, $(t, b; \nu_\tau, \tau^-)$, and their antiparticles, which appear to differ from the first only in their larger masses and their kinematically allowed decay modes, are not obviously needed for anything.[4] Do they play some essential role in nature that we haven't yet perceived? Conversely, are they perhaps some necessary or accidental consequence of a more fundamental unified theory?

Perhaps even more puzzling is the enormous range of fermion masses, ranging from 0.51 MeV for the e^-

---

[4]It is convenient that the known heavy fermions fall in complete family repetitions of the (u, d, ν_e, e^-) because that set of quantum numbers leads to the cancellation of anomalies. However, there are other possible sets of new particles that could emerge in BSM physics that also avoid anomalies.

to 173 GeV for the t quark, compared with the proton mass of 938 MeV. The mixing angles and phase in the CKM matrix V_q defined in equation (4.57) on page 98 are also not understood. These and the mass eigenvalues both result from the original fermion mass matrices, which are themselves proportional to the Higgs-Yukawa couplings, so it is likely that the small quark mixings are associated with the small mass ratios, but the details are unclear.

The neutrinos are even stranger, weighing in at less than around 0.1 eV, and we don't even know whether they are Dirac or Majorana. At least in this case, the small values may be explained by *large* mass scales for underlying new physics, as in the seesaw model, but again the details, including the rationale for two large and one small mixing angle, are unclear.

A related issue involves CP violation, which is necessary to dynamically generate the observed excess of matter with respect to antimatter (baryogenesis). Neither the CP violation associated with the CKM matrix nor any analogous leptonic CP violation is sufficient to account for the observed asymmetry. Some additional source of CP violation, e.g., associated with the heavy Majorana neutrinos in the seesaw model or with new physics during the electroweak phase transition (section 3.3), is needed. This will be further discussed in chapter 6.

There has been enormous theoretical effort to understand the fermion spectrum, mixings, and CP phases, including detailed top-down unification or superstring models, new symmetries relating the families, compositeness, wave functions in extra dimensions of space, and mass generation from higher-order terms in perturbation theory.

I think it is fair to say that none of these approaches has yet met with unqualified success. It is possible that some of these ideas will eventually lead to an understanding of the flavor problem, but it is also possible that we have been asking the wrong questions.

Incredible Fine-Tuning of the Electroweak Scale

It is remarkable that the Higgs mass, $M_H \sim 125$ GeV, and the closely related electroweak scale, $v = M_H/\sqrt{2\lambda} \sim 246$ GeV, are some 17 orders of magnitude smaller than the Planck scale $M_P = G_N^{-1/2} \sim 10^{19}$ GeV. One might expect that if gravity and the electroweak interactions are somehow unified or otherwise related, the two scales should not differ by more than a few orders of magnitude. To be more concrete, the physical (renormalized) Higgs mass-square M_H^2 in the standard model is the sum of the *bare* mass-squared parameter $(M_H^2)_{bare}$ from the Lagrangian density and the higher-order corrections given by diagrams such as those in figure 5.1. These diagrams would diverge quadratically if the internal momenta were allowed to go to infinity, but in practice the momentum integrals are cut off by the new physics scale Λ_{NP}, above which the SM is no longer valid, so that

$$M_H^2 = (M_H^2)_{bare} + \mathcal{O}(\lambda, g_V^2, h_t^2)\Lambda_{NP}^2. \qquad (5.1)$$

If there is no new physics up to the Planck scale, i.e., $\Lambda_{NP} = M_P$, then the higher-order corrections to M_H^2 are around 10^{34} times larger than the physical value! Technically, this is not inconsistent: one can choose $(M_H^2)_{bare}$ so

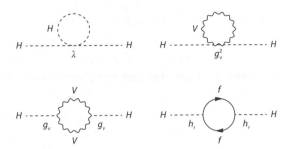

Figure 5.1. One-loop corrections to the square of the Higgs mass in the standard model, from Langacker 2010. V can represent W or Z, with $g_W \equiv g$ and $g_Z \equiv \sqrt{g^2 + g'^2}$. λ is the quartic Higgs self-interaction, and h_f is the Higgs-Yukawa coupling to fermion f. The largest h_f is for the top quark, $h_t = \mathcal{O}(1)$.

that it cancels the corrections to 33 decimal places,[5] but this is hard to swallow. Another way of putting this is that the physics at low energies is incredibly (preposterously?) sensitive[6] to the physics at Λ_{NP}. The need for this incredible fine-tuning of $(M_H^2)_{bare}$ is known as the *Higgs-hierarchy problem*.

The Higgs-hierarchy problem has driven much of the work on possible physics beyond the standard model. One class of proposed solutions involves the introduction of additional particles with couplings related to those of the

---

[5]The inclusion of even higher-order corrections would require the readjustment of the fine-tuning to every order in perturbation theory.

[6]In contrast, the corrections to renormalizable coupling constants and fermion masses are typically not large because they depend logarithmically only on the scale. For example, the relation between the scale $\Lambda_{QCD} \lesssim 1$ GeV at which the strong coupling α_s becomes large and Λ_{NP} is $\ln\left(\frac{\Lambda_{NP}}{\Lambda_{QCD}}\right) \sim \frac{0.9}{\alpha_s(\Lambda_{NP}^2)}$. For $\Lambda_{NP} = M_P$, the observed QCD scale is obtained for $\alpha_s(M_P^2) \sim 0.02$ without any particular fine-tuning.

SM by new symmetries in such a way that they cancel
the quadratic divergences of the diagrams in figure 5.1.
There are typically residual finite corrections associated
with the scale at which the new symmetries are broken,
but as long as that scale is around a TeV or so they are
not uncomfortably large. The most popular example is
supersymmetry, in which each SM particle is accompanied
by a new supersymmetric partner which differs in spin
by $\pm 1/2$. Supersymmetry and its other motivations and
consequences will be described in chapter 6. Here we just
emphasize that supersymmetry is fundamentally different
from the internal symmetries discussed in the previous
chapter in that it relates fermions to bosons. There are
also other classes of nonsupersymmetric theories in which
cancellations are between fermions and between bosons.

Another possibility is that the Higgs boson is not
elementary, in which case Λ_{NP} would be set by the mass
scale of the constituents, presumably in the TeV range.

A third option is that the largest fundamental scale of
physics, M_F, is not M_P but is instead very much lower,
perhaps in the TeV range, so that the corrections to M_H^2
would be cut off at M_F. This could occur if there are
additional dimensions of space, which could alter the relation between the gravitational constant and M_F because
of the spreading of the gravitational field lines into the
extra dimensions, i.e., $M_F \lll G_N^{-1/2} \equiv M_P$. Of course,
we "know" that there are only three space dimensions
because we don't perceive any others. But that is not really
compelling. For example, if an extra space dimension is
curled up in a tiny circle, we might not be aware of its
existence. (Think of an ant crawling along the interior of

a narrow drinking straw—it could move only along the length and might not notice the curved direction.) Even if the extra dimensions are not small, they would be difficult to perceive if we (i.e., ordinary particles, gauge bosons, etc.) are somehow stuck on our three-dimensional surface.

All of these possibilities involve new observable effects at or near the scale of the corrections to M_H^2. In supersymmetry or other "cancellation" models, there are new particles and interactions. If the Higgs is not elementary, there are the constituents, other bound states, and particles associated with the binding. In the extra-dimensional theories, there would be *Kaluza-Klein* excitations of the graviton and of any other particles that can propagate in the extra dimensions. These are similar to the original particles, except they have additional (apparent) masses associated with their quantized momenta in the extra dimensions. For a finite dimension of radius R, these are typically multiples of $1/R$.

These classes of theories are said to be *natural* if the new physics scale is not too large compared to M_H, e.g., $\lesssim 1$ TeV. Naturalness has long been an almost unquestioned assumption in particle physics, leading to considerable optimism that new physics would be found in the initial (7–8 TeV) run of the LHC. However, no new physics was observed. There is still an excellent chance that effects will be seen in the next (13–14 TeV) run, perhaps corresponding to Λ_{NP} of a few TeV. Even this would require some fine-tuning (the *little hierarchy problem*), though it would be far less severe than for $\Lambda_{NP} = M_P$. For these reasons, some physicists are starting to question the naturalness paradigm. Perhaps the new physics scale really is M_P and the necessary "fine-tuning" is really the consequence of a totally different paradigm, such as a vast *landscape* of superstring vacua, with the one we live in determined by *environmental* (*anthropic*) considerations. These ideas will be further discussed in section 5.3.

Time Reversal and the Shape of the Neutron

The Lagrangian density for quantum electrodynamics, equation (4.16) on page 59, could in principle be supplemented by an additional gauge-invariant term

$$\mathcal{L}_{\theta_{QED}} = \frac{\theta_{QED}}{32\pi^2} e^2 F_{\mu\nu} \bar{F}^{\mu\nu}, \qquad (5.2)$$

where θ_{QED} is a dimensionless constant, the $e^2/32\pi^2$ is for convenience, and the *dual field strength tensor* $\bar{F}^{\mu\nu}$ is obtained from (4.11) by interchanging $\vec{E}$ and $\vec{B}$. In fact, $\mathcal{L}_{\theta_{QED}}$ does not affect any physical process,[7] but it is useful for illustration. From (4.11), one has $F_{\mu\nu}\bar{F}^{\mu\nu} = 4\vec{E} \cdot \vec{B}$. Under space reflection, $\vec{E} \to -\vec{E}$ and $\vec{B} \to +\vec{B}$, while $\vec{E} \to +\vec{E}$ and $\vec{B} \to -\vec{B}$ under time reversal. Both reverse sign under charge conjugation. Therefore, $\mathcal{L}_{\theta_{QED}}$ would violate P, T, and CP invariance if it were observable, but would be invariant under C and CPT.

A similar P, T, and CP-violating term,[8]

$$\mathcal{L}_{\theta_{QCD}} = \frac{\theta_{QCD}}{32\pi^2} g_s^2 G_{i\mu\nu} \bar{G}_i^{\mu\nu}, \qquad (5.3)$$

---

[7] $F_{\mu\nu}\bar{F}^{\mu\nu}$ can be written as $\partial^\mu K_\mu$, where K_μ is a current. This does not affect the action, which can be written as the integral of $\mathcal{L}$ over space-time, provided the fields fall off sufficiently rapidly at large x.

[8] $\bar{G}_i^{\mu\nu}$ is defined as $\frac{1}{2}\epsilon_{\mu\nu\rho\sigma} G_i^{\rho\sigma}$, where $\epsilon_{\mu\nu\rho\sigma}$ is the antisymmetric tensor with $\epsilon_{0123} = 1$.

can be added to $\mathcal{L}_{QCD}$ given in (4.30) on page 74. Unlike the QED case, $\mathcal{L}_{\theta_{QCD}}$ is observable.[9] Its most important consequence is that it can generate a nonzero *electric dipole moment* (EDM) $\vec{d}_n$ for the neutron, with magnitude $d_n \sim \theta_{QCD} \times 3 \times 10^{-16}$ e-cm.

A neutron EDM would correspond to a separation between positive and negative charges, i.e., a nonspherical shape, something like in a Na^+Cl^- molecule. $\vec{d}_n$ would have to either align or anti-align with the spin for all neutrons (otherwise there would be an extra degree of freedom). Under space reflection, $\vec{d}_n \to -\vec{d}_n$, while the spin direction would not change. A nonzero d_n would therefore mean that nature is not invariant. Similarly, $\vec{d}_n$ is unchanged but spin reverses direction under time-reversal, so a nonzero value would imply T violation.

There is no evidence for a nonzero dipole moment. The experimental upper limit, $d_n < 2.9 \times 10^{-26}$ e-cm, is extremely stringent, implying $\theta_{QCD} < 10^{-10}$. ($\mathcal{L}_{\theta_{QCD}}$ also induces a proton EDM, but there the experimental constraint is much weaker.) The need for such a small value of the dimensionless θ_{QCD} is the *strong CP problem*. (T and CP noninvariance are essentially the same thing provided that CPT is a good symmetry.)

One might think that it would suffice to simply impose CP invariance, that is, just set $\theta_{QCD} = 0$, but unfortunately that does not do the job. We know that there is CP violation associated with the CKM matrix, ultimately due to phases in the Higgs-Yukawa couplings. It turns out

---

[9] $G_{i\mu\nu}\tilde{G}_i^{\mu\nu}$ can again be written as a four-divergence, but in this case there are nontrivial gauge field configurations that prevent the surface terms in the action from vanishing.

that there is a subtle connection between those phases and θ_{QCD}, associated with an anomaly in the unitary transformations that were introduced on page 95 to diagonalize the quark mass matrices. Those transformations lead to corrections of $\mathcal{O}(10^{-3})$ to θ_{QCD}, which must be delicately cancelled against the bare value. This is not so extreme as the fine-tuning that we encountered in the Higgs-hierarchy problem or the even more difficult cosmological constant problem considered later, but it is still somewhat disturbing, especially since it has no known environmental solution.

It turns out that $\mathcal{L}_{\theta_{QCD}}$ would become unobservable if the Higgs-generated part of one of the quark masses were zero (the coefficient of d_n is proportional to the product of masses). Even though the u quark mass is very small (table 3.2), however, it is difficult to stretch the uncertainties that far. Another possibility is to impose CP invariance on the entire Lagrangian density, so that CP is broken spontaneously. This would be analogous to the spontaneously broken continuous symmetries in section 4.6. The corrections to θ_{QCD} would still be much too large unless additional symmetries are imposed to keep them under control. Moreover, spontaneously broken *discrete* (i.e., not depending on a continuous parameter) symmetries like CP can lead to serious problems with cosmological *domain walls*. Models satisfying all of these criteria exist but are rather complicated.

The most popular solution to the strong CP problem is to extend the SM with an additional global symmetry in such a way that θ_{QCD} becomes a scalar field rather than a parameter (Peccei and Quinn 1977). The physical value of θ_{QCD} then vanishes at the minimum of the potential. Spontaneous breaking leads to a Nambu-Goldstone boson known as the *axion*, with a very tiny mass (typically $\ll 1$ eV) induced by anomaly effects. Axions could also explain the cosmological dark matter, and experimental searches are under way.

Electric dipole moments can also be generated by other sources of CP violation. There is an active experimental program searching for them, with especially stringent limits on the EDMs of the neutron, the electron, and the Hg atom. In the SM, the EDMs due to the CP violation in the CKM matrix (with the exception of those associated with the corrections to θ_{QCD}) are of high order and are predicted to be much smaller than the experimental limits. Those from the leptonic mixing should be even smaller. However, most extensions of the SM involve new sources of CP violation that can lead to much larger EDMs than those in the SM. The current and future EDM searches have significantly constrained such models, especially for new physics at the TeV scale. Similar statements apply to flavor changing neutral currents.

Quantum Gravity and the Cosmological Constant

Classical general relativity can be included in the SM, but it has no direct connection to the other interactions. Moreover, it is not quantum mechanical, and straightforward attempts to quantize it lead to nonrenormalizable theories with severe divergences.

The local (gauge) version of supersymmetry (*supergravity*) connects gravity more naturally to the other interactions, but does not tackle the renormalizability. Superstring theory, on the other hand, not only requires the existence of quantum gravity but also makes it and the other interactions *finite*, not just renormalizable. More in chapter 6.

Perhaps the most difficult aspect of gravity is that in the presence of spontaneous symmetry breaking (SSB) the minimum of the Higgs potential has a nonzero constant value, given by the first term in equation (4.39) on page 89. A constant of the energy has no effect on microscopic physics or Newtonian gravity, but it does couple to gravity in general relativity, where it is known as a *cosmological constant*, $\Lambda \equiv 8\pi G_N \langle 0|V|0\rangle$.

A possible cosmological constant has had a long history. Einstein included Λ in his original formulation of general relativity to allow for a static Universe (although it turns out that the static solution is unstable). He subsequently abandoned it after Hubble observed that the Universe is expanding, and he was reputed to have said that the cosmological constant was the biggest blunder of his life. Nevertheless, the theoretical possibility of Λ reemerged in field theory, as a possible consequence of SSB or of the zero-point energy associated with the harmonic oscillator-like momentum modes of a quantum field. The subject came full circle with the observation that the Universe is accelerating, as briefly described at the end of section 3.3. The dark energy that drives the acceleration could be a cosmological constant or something very similar.

Unfortunately, the vacuum energy associated with the Higgs potential is vastly too large in magnitude (by a factor $\sim 10^{54}$) to be the observed dark energy. It is also of the wrong sign, and would lead to deceleration and collapse rather than acceleration. Similar to the Higgs-hierarchy and strong CP problems, the cosmological constant induced by spontaneous symmetry breaking could be cancelled by some "primordial" vacuum energy. This would be equivalent to adding a constant to the original Higgs potential. For many years, most physicists thought that there must be some not yet understood mechanism that would force the sum of the primordial and SSB-induced terms to cancel. However, the realization that Λ is nonzero seems to make matters worse: it is hard to understand why the sum of two apparently unrelated terms should almost but not quite exactly cancel.[10]

Unfortunately, superstrings do not solve the cosmological constant problem, but seem to greatly aggravate it: most string vacua lead to cosmological constants around 123 orders of magnitude too large! Grand unified theories are almost as problematic. Many consider the cosmological constant to be the most serious problem in particle physics (Weinberg 1989). Perhaps the most plausible explanation is environmental.

5.2 Terra Incognita: Unanswered Questions

There are a number of issues that are not really addressed by the standard model, or for which the SM is perhaps *too* successful.

---

[10]There are additional much smaller contributions from the QCD vacuum that are still enormous compared to the observed value and would also have to be cancelled.

The Matter (Baryon) Asymmetry

It is straightforward to produce antimatter in the laboratory, and, to a good approximation, matter and antimatter
have similar properties. However, it is fortunate for our
existence that there isn't much antimatter around us[11]—
if there were, the matter and antimatter would quickly
annihilate. This asymmetry is not just local: if there were
distant antimatter stars, galaxies, or clusters of galaxies,
we would observe photons created by annihilations at
the boundaries between the matter and antimatter. Any
antimatter regions would have to be separated by nearly
the size of the observable Universe (Canetti et al. 2012). As
mentioned in section 3.3, the very early Universe involved
a hot plasma that included quarks, antiquarks, photons,
and other particles. One can work backward from the ratio
of baryons to photons observed today (or, more precisely,
in the CMB and BBN) to infer that at some very early
stage there must have been a tiny excess of around one part
in 10^9 of quarks with repect to antiquarks (the matter-
antimatter or baryon asymmetry). The antiquarks later
annihilated, with the residual excess of quarks responsible
for the ordinary matter that we are made of. The approximate electrical neutrality of the Universe implies a similar
nonzero but tiny excess of electrons over positrons. There
are no stringent constraints on the neutrinos, but it is likely
that any $\nu - \bar{\nu}$ asymmetry is comparable.

How did this $\frac{n_q - n_{\bar{q}}}{n_q} \sim 10^{-9}$ asymmetry come about? It
could in principle have been there as an initial condition

---

[11]There are small amounts of antimatter produced in radioactive decays,
cosmic ray interactions, other astrophysical events, and high-energy physics labs.

on the big bang, but this is excluded if the Universe underwent an initial period of rapid inflation or similar solution to the flatness problem (section 3.3), which would have diluted the particle densities to negligible amounts. In that case, it must have somehow been produced dynamically following reheating. In 1967, Andrei Sakharov (later famous as a Soviet dissident and human rights advocate) published a prescient study (Sakharov 1967) of the ingredients needed for this baryogenesis. They were

1. The nonconservation of baryon number B (the number of baryons minus antibaryons).
2. CP and C violation, so that more quarks than antiquarks could be produced.
3. Nonequilibrium of the B-violating processes. Otherwise, the number of quarks and antiquarks would have to be equal since they are degenerate by CPT. (An alternative would be to allow CPT violation.)

The standard model does violate CP. There are also tiny and subtle B and L violating effects associated with tunneling between one electroweak vacuum gauge field configuration and another. The latter are nonperturbative and negligibly small {of $\mathcal{O}(\exp[-2\pi \sin^2 \theta_W/\alpha] \sim 10^{-80})$} at present, but could have been important in the early Universe due to thermal fluctuations between configuration of different $B + L$. Unfortunately, the CP violation from the CKM and PMNS matrices in the SM is too small to be relevant to baryogenesis, and the electroweak phase transition, after which the Higgs developed its nonzero VEV, was too gradual to satisfy the third condition.

One must therefore go beyond the standard model (and probably beyond the simplest supersymmetric extension). There are many speculative models of BSM physics that could account for the baryon asymmetry. We will mention two possibilities. One is *electroweak baryogenesis* (EWBG), in which the nature of the electroweak phase transition is modified, e.g., by a more complicated Higgs structure, in such a way that it is strongly first order. That means that as the Universe cooled in the unbroken phase, at some critical temperature bubbles with the nonzero Higgs VEV inside nucleated and expanded, eventually filling space. This nonequilibrium process, combined with some new source of CP violation associated with the bubble and the high-temperature $B + L$ violation, could possibly have generated the asymmetry. A necessary consequence of EWBG is that there should be new physics associated with the phase transition and new sources of CP violation at or not too far above the TeV scale, both of which might be directly observable in the laboratory.

Another possibility is leptogenesis, associated with the seesaw models for small neutrino mass mentioned on page 136. At very high temperature in the early Universe the heavy Majorana neutrinos ν_R would have been in equilibrium. However, for $T \ll M$, they could no longer stay in equilibrium and would eventually decay. The Majorana neutrino mass violates lepton number (L), so ν_R could decay into either a lepton or an antilepton, e.g., into $e^-\phi^+$ or $e^+\phi^-$, where $\phi^\pm$ are the charged Higgs particles relevant at high temperatures where $SU(2) \times U(1)$ is unbroken. New CP phases associated with the heavy neutrinos could lead to unequal rates, i.e., to an excess of antileptons over leptons. Some of this excess could later be converted to a baryon excess by nonperturbative electroweak vacuum effects. This scenario is appealing in that it does not require any new physics beyond the seesaw model, but it has the disadvantage that most versions do not lead to any directly observable effects at the TeV scale.

Dark Matter and Energy

Astronomers have repeatedly demoted humankind's place in the Universe. First was the realization that the Earth revolves around the Sun and not the other way around. Furthermore, the Sun is not unique or at the center: it is just a fairly typical star, similar to those that we see in the night sky, many of which have their own planetary systems. Even the numerous stars visible to the naked eye are not the whole story. They are a small subset of the more than 10^{11} in our Milky Way galaxy. Moreover, there are a myriad ($\gtrsim 10^{11}$) of other galaxies, stretching out as far as we can see with the largest telescopes.

More recently, it has become apparent that the atoms and molecules that we are made of constitute only around 5% of the stuff in the Universe. As described at the end of section 3.3 the remainder is the mysterious dark energy (70%) and dark matter (25%).

Dark energy refers to energy that does not change in density as the Universe expands, such as the energy of the vacuum itself. If positive, it acts as a repulsive gravity, leading to acceleration of the expansion. The dark energy could be a cosmological constant associated with the energy stored in the potential of a scalar field when evaluated at its minimum, or it could be due to a slowly varying field (quintessence). As detailed in section 5.1, the observed dark energy is many orders of magnitude smaller than the expected value from the VEV of the SM Higgs field unless there is a fine-tuned cancellation of terms.

The origin, sign, and magnitude of the dark energy are a mystery. Is it somehow related to a much larger energy density that led to a brief instant of inflation in the very early Universe (and which is itself not understood)? If it involves a slowly varying field, there could be a time (or space) dependence of coupling "constants" (e.g., Uzan 2003). For example, the induced cubic Higgs self-interaction in (4.39) on page 89 or the VVH interactions in (4.48) are proportional to the Higgs VEV v and could therefore change if v varied with time. Other couplings, such as gauge couplings, could also be ultimately associated with the values of scalar fields (this often occurs in superstring theories). There are stringent laboratory limits on time variation of the fine structure constant, for example, but it is conceivable that there have been small changes (e.g., of order one part in 10^5) on the timescale of the age of the Universe.

Both BBN and the CMB have established that the dark matter is really something new, i.e., it is not ordinary matter hidden in some form (such as very faint stars) that has not been detected. Neutrino masses in the 10 eV range could have provided enough dark matter, but do not lead to the observed distribution of galaxy sizes or CMB properties (see page 133). Rather than a new form of matter, the original observations could possibly have been accounted for by modifying the large distance behavior of gravity (*Modified Newtonian Dynamics*, or *MOND*), but observations of colliding galaxies, in which the ordinary and dark matter become separated, make this unlikely.

Most likely, the dark matter consists of some type of (BSM) elementary particle that interacts only weakly with ordinary matter and radiation, and which is either absolutely stable or at least has a lifetime longer than (or comparable to) the 14 billion year age of the Universe. There are many theoretical candidates for such particles (e.g., Gelmini 2015). The most popular are the spin-1/2 partners of the neutral Higgs and gauge bosons in supersymmetry (*WIMPs*), or the very light axions that were motivated by the strong CP problem. They could also be associated with an entirely new *dark sector* of particles and interactions that is only weakly coupled to our sector of physics. Also possible are *primordial black holes* with masses around $10^{-10} M_\odot$, which could have been produced in a phase transition or other dramatic event in the very early Universe (Green 2015).

The nature of the dark matter and dark energy is one of the most intriguing issues in physics. A possibly relevant fact is that these and the ordinary matter are all within an order of magnitude or so in their contribution the total energy density of the present Universe. Is this just a coincidence? They could have been drastically different.

Rare Processes and the Stability of Matter

In some ways, the standard model seems *too* successful. For example, the strong suppression of flavor-changing

neutral current (FCNC) effects in the kaon and lepton
systems (section 4.7) is an automatic consequence of the
GIM mechanism, the minimal Higgs structure, and the
small neutrino masses. Similarly, electric dipole moments
(EDMs) occur only at high order (except for those associated with strong CP violation) and are predicted to
be negligible.

The suppression of FCNCs and EDMs in the SM can
be thought of as accidental. That is, they are not forbidden
by the SM gauge symmetries, but rather are forced to
be of higher order and small because of the SM particle
content. This is no longer the case in general: most of
the TeV-scale BSM theories proposed to solve the Higgs-
hierarchy problem have new sources of both FCNCs
and CP violation that are potentially in conflict with
observations. This has led to some models being discarded,
while others are constrained to small or fine-tuned regions
of parameters, or to other ways to minimize the effects. The
nonobservation of new physics in the first LHC run pushes
the possible BSM mass scales higher. This aggravates
the Higgs-hierarchy and naturalness problems, but also
somewhat relaxes the FCNC and EDM constraints.

The stability of matter is especially dramatic. The SM
gauge symmetries would allow the proton to decay, for
example $p \rightarrow e^+\pi^0$. Similarly, bound neutrons stable
against β decay (because of nuclear binding energy) could
decay into $e^+\pi^-$. Fortunately, this does not occur rapidly,
or we would not be here. The distinguished experimental
physicist Maurice Goldhaber used to quip something like
"we could feel any rapid decay in our bones," meaning
that we would quickly acquire cancer from the proton

decay products if the lifetime were less than 10^{16} yr. Early experiments established a more stringent lower limit of $\sim 10^{20}$ yr, while more recent dedicated searches have established that the lifetimes for decays into prominent modes[12] exceed 10^{31}–10^{34} yr. For example, the SuperKamiokande experiment, described on page 132, found $\tau > 8 \times 10^{33}$ yr for the $p \to e^+ \pi^0$ lifetime.

Prior to the SM, one simply postulated the existence of a conserved baryon number B to ensure proton stability. The SM, however, has the elegant feature that this is automatic: baryon and lepton number (L) are global symmetries at the renormalizable level because there is no way to write renormalizable B or L-violating interactions from the available fields (Weinberg 1980b).

Of course, the matter asymmetry strongly suggests that there is B violation in nature—it just has to be very small. We already mentioned the nonperturbative vacuum-tunneling effects in the SM, which violate $B + L$ but are negligibly small except at very high termperatures. Many extensions of the SM allow perturbative B and/or L violation associated with the exchange of new heavy particles. L-violating effects could induce the operator in (4.69) on page 135, leading to small Majorana neutrino masses. However, B violation (or more frequently, $B + L$ violation), which occurs in grand unified theories and many superstring theories, is much more dangerous because it can lead to $p \to e^+ \pi^0$, $\bar{\nu} K^+$, or other decay modes. One must typically require that the relevant mass

---

[12] The lifetime must exceed 10^{29} yr even if the decay products are not directly detectable, because of the subsequent nuclear transitions that could be observed following the disappearance of a bound proton or neutron.

scale of the new particles is very large,[13] e.g., $> 10^{16}$ GeV. It is interesting that similar scales are suggested by the possible unification of gauge couplings.

Is Nature Just Right?

Does nature need to be so complicated as the standard model? I certainly will not attempt to give a definitive answer, but it is interesting to examine the essential roles played by the variety of interactions and particles, and also by the parameters. This will be relevant to the discussion of such paradigms as naturalness, uniqueness, and minimality in the next section. It is also interesting for its own sake to think about whether something as complex as life could have existed if the rules had been a little different.

In fact, each of the known interactions does appear to be "essential" for life, in the (limited) sense that perturbing around our physics by simply eliminating any one of them would have catastrophic consequences. Gravity is necessary to "bring things together" into galaxies, stars, planets, etc., so that something can happen in a sparse universe. Similarly, a variety of types of atoms and molecules (or something analogous) are needed for chemistry, materials, and life as we know it. Their existence and associated chemical reactions require something like loosely bound electrons and electromagnetism. A variety of atoms also requires many types of nuclei, which occur in our nature by starting with only two kinds of stable or quasi-stable

---

[13]In some versions of supersymmetry, proton decay can be mediated by supersymmetric partners. Additional symmetries must be imposed to forbid these couplings if the supersymmetry scale is in the TeV range.

nucleons. These are held together and stabilized in various numbers and arrangements in a delicate balance between the strength and range of the strong interaction and of electromagnetism, quantum dynamics, the Pauli exclusion principle, and the proton and neutron masses. The strong interaction, which is fundamentally due to QCD and the bound states it produces, is also needed for energy generation in stars. This prevents stellar collapse and is ultimately responsible for most of our energy needs. Thus, gravity, electromagnetism, and the strong interactions hold things together, while chemical and nuclear reactions involve rearrangements of electrons or nucleons, respectively.

The weak interactions do not hold anything together. Rather, they allow transformations between protons and neutrons, essential for energy generation in stars and for the nucleosynthesis of elements heavier than hydrogen, mainly in the early universe (especially for ^4He), stars (elements up to ^{56}Fe), and in core-collapse supernovae (for still heavier unstable elements).

Each of the known interactions therefore plays an essential role in the creation and existence of complex structures and life, at least in the sense described above. What about the fundamental (point-like) particles? The e^-, ν_e, and the three colors of u and d quarks all have necessary roles in our scenario as well. Furthermore, the existence of a variety of atoms and nuclei depends critically on the masses of these particles relative to each other and relative to the strength of electromagnetism. For example, if $m_d - m_u$ were increased sufficiently, then the neutron would become heavier relative to the proton, and nuclei other than the proton would be unstable because of the more rapid neutron β decay. Conversely, a smaller or negative $m_d - m_u$ (or a heavier electron) could destabilize even the hydrogen atom by e^- capture (and possibly $p \to ne^+\nu_e$). The sum $m_d + m_u$ is also constrained by nuclear binding effects (for example, the pion mass depends sensitively on $m_d + m_u$, as mentioned in the note on page 90). These considerations lead to fairly narrow allowed regions for m_d, m_u, and m_e, and therefore on the Higgs-Yukawa couplings and Higgs VEV, assuming that everything else in the SM is unchanged (Damour and Donoghue 2008).

Other such lucky coincidences are reviewed in Schellekens 2013. The most famous example concerns the dark energy or cosmological constant (Weinberg 1989). We have seen that the dark energy associated with the Higgs VEV is more than 50 orders of magnitude larger than the observed value, and of the wrong sign, while the value naïvely expected from superstring theory or other fundamental theories of gravity is larger by another 60 orders! We are indeed fortunate that the true value is so tiny, perhaps due to delicate cancellations. A positive dark energy density much larger than the observed one would have led to such a rapidly (exponentially) expanding Universe that structures like galaxies or stars or people could never have formed, and the Universe would be virtually empty. Conversely, a large negative value would have caused the Universe to recollapse into an extremely hot and dense *big crunch* long before life could have developed.

Therefore, many aspects of the standard model are in some sense "essential." It must be strongly emphasized, however, that there are probably many other scenarios that could do just as well. These could involve correlated changes in many of the features and/or parameters of the SM, or in the cosmological history. Other possibilities might be completely different from the SM. A systematic study is probably beyond our capabilities.

Not everything in the SM appears to be necessary, however, even in the limited sense of the consequences of varying one aspect at a time. In particular, the first-family particles (u, d; e^-, ν_e) are all that are needed for matter under ordinary terrestrial conditions. We are still ignorant as to why nature has two heavier families, as mentioned on page 140.

5.3 Are the Paradigms Correct?

In addition to the ingredients needed to describe a physical system mentioned in section 2.1, there are several other issues that are often ignored or taken for granted. Recently, however, the lack of clear experimental evidence for physics beyond the standard model has challenged the paradigm of naturalness, while some theoretical notions question uniqueness and minimality.

Our discussion of these paradigms takes us somewhat outside of the traditional descriptive domain of physics and more into that of philosophy. However, they are interesting in themselves and are useful in guiding our considerations of what might underlie the standard model. The time may be ripe.

Naturalness or Tuning?

Naturalness in physics refers to the assumption that the qualitative properties of nature should not depend on a fine-tuning of the inputs, e.g., that dimensionless ratios should not be extremely large or small without a good reason, or that unrelated quantities should not have to cancel to a high precision. In the past, naturalness has frequently been a guide to the existence of new physics (e.g., Giudice 2008).

As an example, consider the contribution of electromagnetism to the mass of the electron. For a classical electron of charge $-e$ distributed uniformly throughout a sphere of radius a, the rest mass is

$$m_e = m_e^0 + \frac{3}{5}\frac{\alpha}{a}, \qquad (5.4)$$

where m_e^0 is the bare or nonelectromagnetic part and $\alpha = e^2/4\pi$. The second term is the electrostatic energy, which diverges if the electron is point-like ($a = 0$). Even using the limit $a < 10^{-17}$ cm from scattering experiments, the Coulomb contribution is more than 10^4 times the actual mass of 0.5 MeV, requiring an unnatural fine-tuned cancellation with m_e^0. The situation is aggravated if one includes the magnetic energy for a spinning electron.

The problem is also made worse by going to a quantum theory involving only photons and relativistic electrons. The perturbative contribution to the electron mass diverges quadratically, i.e., is of $\mathcal{O}(\alpha\Lambda^2)$, where Λ is the energy scale at which the sum over intermediate states is cut off. Equivalently, it is of $\mathcal{O}(\alpha/a^2)$, where $a \equiv 1/\Lambda$ is the corresponding distance scale. A fine-tuned cancellation with the bare mass would again be required.

Naturalness is saved when one takes into account the existence of the positron, which is required by the union of quantum mechanics and relativity in the Dirac equation: the quadratic divergence is cancelled by intermediate states that involve positrons as well as electrons and photons,[14] leaving only relatively innocuous logarithmic divergences. The one-loop QED calculation yields

$$m_e = m_e^0 \left(1 + \frac{6\alpha}{4\pi} \ln \frac{\Lambda}{m_e} \right). \qquad (5.5)$$

The correction is proportional to the bare mass, and even for $\Lambda = M_P$ the coefficient is only ~ 0.18, so no unnatural fine-tuned cancellation is required.[15] In an alternate history, the positron could have been predicted by the assumption of naturalness prior to its actual discovery in 1932!

As another example, the dominant contribution[16] to the $\pi^+ - \pi^0$ mass difference, $m_{\pi^+} - m_{\pi^0} = 4.6$ MeV, is electromagnetic. However, if one treats the pions as point-like particles, the one-photon contribution diverges

---

[14]These are incorporated automatically in the Feynman diagram formulation of perturbation theory.

[15]The small value of $m_e^0 \sim h_e v$ must still be explained. $v \lll M_P$ and $h_e \sim 10^{-5}$ are respectively just the Higgs-hierarchy problem and (part of) the flavor problem described in section 5.1.

[16]Exact isospin invariance of the strong interactions would not allow a bare mass difference, while the known isospin breaking from $m_u - m_d$ is of higher order and small.

quadratically, similar to that of the Higgs mass-squared in (5.1):

$$m_{\pi^+}^2 - m_{\pi^0}^2 = \frac{3\alpha}{4\pi} \Lambda^2, \qquad (5.6)$$

which corresponds to the observed splitting for $\Lambda \sim 850$ MeV. Naturalness therefore requires that the theory must be modified at some new physics scale $\Lambda_{NP} \lesssim 850$ MeV. Indeed, there are hadronic resonances, such as the vector ρ at 770 MeV and the axial A_1 at 1260 MeV that modify the calculation in agreement with experiment. An equivalent interpretation is that the pions and resonances are really $\bar{q}q$ bound states, with the compositeness scale set by the the QCD-induced quark masses $\sim m_p/3 \gtrsim 300$ MeV.

We finally mention the K_L-K_S mass difference, discussed following equation (4.62) on page 111. With only the u, d, and s quarks, the diagrams in figure 4.13 yielded a finite contribution thousands of times larger than observation. However, naturalness was saved by the GIM mechanism (the existence of the c quark), which predicted an exact (*not* fine-tuned) cancellation between the u and c quark diagrams in the limit $m_c = m_u$. The small residual corrections led to the correct prediction of m_c around 1.5 GeV.

These examples illustrate the remarkable past success of naturalness: fine-tuning was avoided by the existence of new particles or principles, although physicists at the time were not always thinking in those terms.

We discussed in section 5.1 that the standard model has several fine-tuning problems. The Higgs-hierarchy problem has strongly motivated the possibility of new physics at the TeV scale, such as new particles, compositeness or other strong dynamics, or a low fundamental scale associated with additional space dimensions. However, as of late 2015 no evidence for any of these has been observed at the LHC or elsewhere. Similarly, the strong CP problem may find its solution in new symmetries, although experimental confirmation may be difficult. The most dramatic fine-tuning concerns the dark energy or cosmological constant, which appears to be incredibly tiny compared to the naïve expectation from electroweak breaking or a higher-scale unification. Unfortunately, naturalness considerations have so far not yielded any very compelling solutions.

Naturalness may still prevail, and especially in the case of the Higgs-hierarchy problem experimental evidence for relevant new physics may emerge soon at the LHC or elsewhere. Nevertheless, many physicists are starting to think about other possibilities. Environmental selection, which we turn to next, is perhaps the most promising alternative, especially for the cosmological constant and Higgs-hierarchy problems. The most interesting question may involve the paradigm of naturalness itself, and whether it will continue to survive and serve as a guide to the unknown.

Uniqueness or Environment?

Perhaps the most striking insight of the standard model is that the strong, weak, and electromagnetic interactions are all associated with gauge symmetries. Gauge-invariant theories are very much constrained in structure, requiring the existence of spin-1 gauge bosons and determining the form of their interactions with fermions, spin-0 particles, and each other. Nevertheless, they are not entirely unique: the gauge groups, representations, and coupling constants must be specified, and possible spontaneous symmetry breaking usually depends on other aspects of the theory.

On the other hand, the Higgs-Yukawa parameters in the SM are completely arbitrary. Attempts to understand them require new symmetries or principles beyond the standard model, such as additional and often ad hoc symmetries (which may be gauge, global, or discrete), constraints from string theory, compositeness, or extra dimensions.

Most physicists take for granted the notion that there is some unique (and hopefully simple and elegant) theory of nature, and most theoretical work is directed toward constructing models or postulating symmetries or dynamics in search of it. However, the uniqueness paradigm leaves unanswered two fundamental questions:

- Why does nature choose this one possibility out of the infinite number of possible field theories? Is there some new selection principle that nobody has thought of? Perhaps there is some constraint from self-consistency, but so far there is no hint of this. Or maybe it is the simplest and/or most beautiful possible theory. However, the standard model itself is far from simple, and of course beauty is in the eye of the beholder.
- Why is this unique theory so cleverly arranged as to allow the evolution of complex structure, as discussed

in the preceding section? It is not obvious that any of the possible explanations for uniqueness mentioned earlier would simultaneously address this question. Is it just a lucky coincidence?

The uniqueness hypothesis may well turn out to be correct, and hopefully we will ultimately not only find the unique theory but also have satisfactory answers to these questions. However, there is another plausible paradigm that should also be considered, i.e., that of the environment.

Let us begin with an analogy. The great astronomer Johannes Kepler is best known today for his three laws of planetary motion, which he deduced empirically from the detailed observations of Tycho Brahe, and which would later be justified by Newton's laws of gravity and mechanics. However, prior to that work Kepler had spent much time trying to find a fundamental geometric explanation for the relative radii of the orbits of the six known planets. He proposed in his *Mysterium Cosmographicum* (1596) that the planetary radii could be understood in terms of the nesting of the five Platonic solids and their inscribed and circumscribed spheres inside of each other (figure 5.2). Kepler's hypothesis was in agreement with the observations to the precision that was then available, but we now understand that Kepler was addressing the wrong question and that the success of his theory was fortuitous. Many stars have planetary systems, and the details of their orbits depend on the initial conditions of the formation of the stellar system and on its subsequent dynamical evolution, not on a beautiful geometric analogy.

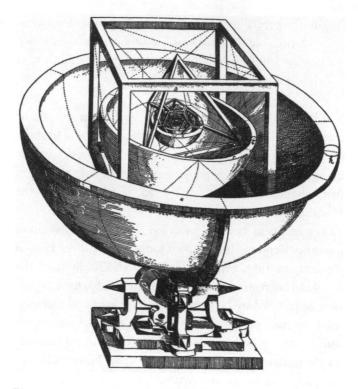

Figure 5.2. Kepler's model of the Solar System, from *Mysterium Cosmographicum.*

Closely related is the question of why the Earth is so suitable for life. Life as we know it requires liquid water, which is most easily accomplished on a planet in the habitable (or Goldilocks) zone that is neither too close to its star nor too distant.[17] However, most of us do not wonder why we are so lucky that the Earth is in the Sun's

---

[17]Liquid water could also occur elsewhere if there are other energy sources, such as tidal heating.

habitable zone; that is the wrong question. Rather, there are almost certainly enormous numbers of stellar systems with appropriate planets, and presumably life is most likely to evolve on the favorable ones.

The environmental paradigm is that there may be no simple unique explanation for some or all of the parameters and features of nature. It instead postulates that there may be an enormous landscape of possible laws of physics, such as might be associated with a fundamental theory with a very large number of vacua or solutions and no underlying selection principle. Even if we assume that each such vacuum is controlled by a field theory, the gauge groups, representations, couplings, geometries of space, and other properties might differ from one to another. For example, it has been estimated that there may be $\gtrsim 10^{600}$ such vacua in string theory, as will be further discussed in chapter 6.

If indeed there is a landscape, then some of these vacua might be "just right" for complex structures and life in the sense discussed starting on page 160, while most would not be. Perhaps some of the fine-tunings and lucky accidents are due to environmental selection, a.k.a. the anthropic principle: in analogy to the lucky accident that the Earth is in the Sun's habitable zone, it could be that if nature were not just right we would not be around to wonder about it. This need not mean that *everything* is environmentally selected. Other features of physics could have conventional explanations (e.g., in symmetries), or just be random (such as the mass of the *b* quark?).

Many years ago, I viewed the anthropic principle as little more than a silly tautology. However, developments in string theory and cosmology have caused me to reconsider. It can be elevated to a plausible scientific possibility if three conditions are satisfied:

- There must be a credible theory that predicts an enormous landscape of possibilities. Superstring theory is an example of a theory that may satisfy this criterion.
- There must be some mechanism for many or all of these vacua to actually be sampled. Fortunately, most versions of inflation (section 3.3) are *eternal*, i.e., localized field fluctuations are constantly producing bubbles that themselves may grow into new regions with different vacua. Most of these would either recollapse immediately or would grow so rapidly that nothing like stars or galaxies could form, but some would be "just right." The most popular version of these ideas is the multiverse: that our observable Universe is just one bubble in an infinite domain of regions,[18] only a tiny fraction of which can support life. Depending on one's point of view, this could be an incredibly exciting expansion of our outlook, or it could represent another step (beyond those mentioned on page 155) in the demotion of humankind's place in nature.
- Any such theory should be verifiable (and falsifiable), at least in part. This is by far the biggest problem and weakest link in the landscape/multiverse ideas, and is a major reason that many physicists are skeptical. Clearly, many aspects will never be amenable to direct

[18]Instead of spatially separated regions, the physics could differ at other times (e.g., if the Universe repeatedly collapses and then reexpands), in spatially overlapping but noninteracting parallel Universes, or in different branches of a quantum wave function.

experimentation or observation. However, I am not completely pessimistic. For example, there could be subtle distortions in the CMB if our region is near one of the boundaries, although no such signals have shown up as yet. Moreover, it is not essential for all aspects of a theory to be tested directly. There is a possibility that eternal inflation and superstring theory will eventually be verified, at least in part (although many physicists are skeptical here as well, especially concerning string theory), and that it will be convincingly shown that they indeed lead to the landscape. In that optimistic scenario we could be reasonably certain that at least the broad outlines of the picture are correct.

Finally, I would argue that the ideas of environmental selection and multiverse should be seriously entertained for the simple reason that they may well be correct. Even if the time is not ripe for directly testing them now, completely new possibilities may emerge in the future. Scientists are clever. Who would have foreseen 60 years ago the role of the CMB in probing the big bang, or who would have guessed even 20 years ago the revolution in paleoanthropology brought about by DNA studies?

Some of these ideas will be touched on again in the remaining chapters.

Minimality or Remnants?

Yet another commonly assumed paradigm is *minimality*, which assumes that nature, or a theory invoked to explain a set of observations, should be as simple as possible. A beautiful example is the quark model, which is far simpler than assuming that all of the hadrons are fundamental. Minimality is often invoked in constructing BSM theories for explaining shortcomings of the standard model.

Minimality is similar in spirit to Occam's razor and certainly seems plausible. However, it is basically an aesthetic issue and should not be considered inviolate. For example, the existence of heavy families does not appear to be minimal, at least with our current understanding, and the standard model itself is far from minimal if one ignores the "just right" considerations. Perhaps more relevant is the fact that some fundamental theories, such as many superstring vacua, often predict new particles or interactions in the low-energy theory that are not needed to solve any particular SM problem. Rather, such remnants managed to accidentally avoid the symmetry-breaking or other mechanisms that led to Planck-scale masses for most of their cousins. They are "just there."

6

HOW WILL WE FIND OUT?

6.1 The Ideas

There are many promising ideas for possible physics beyond the standard model (BSM), some of which have been mentioned in chapter 5. Some are top-down, i.e., are motivated by fundamental theoretical and aesthetic considerations, such as the unification of the interactions. The most popular of these are superstring theory and grand unification, most versions of which involve a very high underlying mass scale, e.g., 10^{16}–10^{19} GeV. Other (bottom-up) ideas are more motivated by experimental considerations or SM problems, and usually postulate new physics at the TeV scale. These categories are not mutually exclusive. Top-down theories often address problems of the SM, while bottom-up models may eventually lead to an ultimate unification. Some ideas, such as supersymmetry, fit into both categories.

Most of the theoretical ideas fall into one or more broad classes:

- New symmetries
- A new layer to the onion

- Extra dimensions
- Dark or hidden sectors
- Unification
- Here be dragons

Symmetries: Super and Otherwise

In chapter 4, we discussed internal symmetries, in which the equations of motion are unchanged under phase changes or rotations of fields into each other. These lead to conserved charges, to the grouping of particles into multiplets with equal masses, and to relations between their interactions. The transformations involve particles with the same spin, and the operators associated with the symmetry transformations obey *commutation* relations, such as (4.23) on page 67. A supersymmetry,[1] on the other hand, involves anticommutation relations and relates the masses and interactions of particles that differ in spin by $1/2$, such as spin-0 and spin-1/2, or spin-1/2 and spin-1. Each fermion must have a bosonic partner, and vice versa.

None of the SM particles have suitable properties to be each other's partners, so the *minimal supersymmetric extension of the standard model* (MSSM) requires a new complex scalar for each left- or right-chiral SM fermion, and a new chiral fermion for each spin-0 or spin-1 boson. The supersymmetry also requires a second Higgs doublet and its spin-1/2 analogs, so the number of "fundamental" fields is more than doubled. The *superpartners* are given

[1] For introductions, see, e.g., Baer and Tata 2006; Olive et al. 2014; Dine 2015.

rather whimsical names: the spin-0 partners of the quarks and leptons are respectively the *squarks* and *sleptons*; the spin-1/2 partner of the gluon is the *gluino*; while the charged and neutral partners of the Higgs scalars and electroweak gauge bosons are the *charginos* and *neutralinos*, respectively. Since none of the new particles have (yet) been observed, they must be very massive, i.e., the supersymmetry must be broken,[2] presumably by *soft* terms like masses that don't mess up what supersymmetry is supposed to accomplish. Such terms ultimately set the weak interaction scale. Unfortunately, the most general soft supersymmetry breaking involves more than 100 free parameters! For this and other reasons, the MSSM is really a whole class of theories.

Given these complications, it perhaps seems preposterous to even consider supersymmetry. Fortunately, there are also a number of advantages.

- It provides a solution to the Higgs-hierarchy problem (page 142). The diagrams involving the superpartners cancel the quadratic divergences in (5.1), with the corrections from (soft) supersymmetry breaking manageable if the superpartner masses are in the TeV range.
- Gauge unification refers to the possibility that the running $SU(3) \times SU(2) \times U(1)$ gauge couplings all meet at some large scale M_X, above which the interactions are all unified. This can be tested using the

---

[2]In the MSSM, the Higgs superpartners can acquire mass independent of supersymmetry breaking. In realistic versions, that mass must be comparable to the soft-breaking masses within an order of magnitude or so. This is ad hoc in the MSSM but can emerge naturally in more general models.

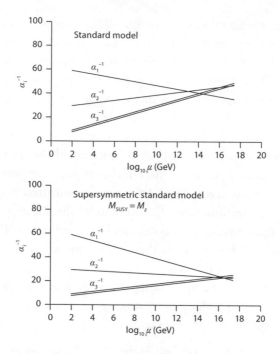

Figure 6.1. The running gauge couplings in the SM (top) and MSSM (bottom), from Langacker 2010. $\alpha_3 = \frac{g_s^2}{4\pi}$ is the QCD coupling, while $\alpha_2 = \frac{g^2}{4\pi}$ and $\alpha_1 = \frac{5}{3}\frac{g'^2}{4\pi}$ are the $SU(2) \times U(1)$ couplings, with 5/3 an appropriate normalization in simple unified theories. The MSSM curves assume that all new particles have masses $M_{SUSY} = M_Z$. A more detailed analysis shows that the agreement is even better if they are at the multi-TeV scale.

observed values at M_Z and the theoretically calculated running. As can be seen in figure 6.1, the gauge couplings almost but do not quite unify assuming that the SM holds to a large scale, but are consistent with unification in the MSSM with $M_X \sim$ few $\times$ 10^{16} GeV.

- The most popular versions of the MSSM involve a new discrete symmetry that differentiates the ordinary particles from the superpartners. That implies that the superpartners can be produced only in pairs, and that their decay products must include an odd number of lighter superpartners. The *lightest supersymmetric partner* (LSP) in those versions must therefore be absolutely stable. It turns out that the lightest neutralino could be a viable dark matter candidate. Neutralinos would have been produced abundantly in the early Universe, and although many would have subsequently annihilated in pairs, their relatively weak interactions would have allowed some to survive to today. The predicted dark matter density for a TeV-scale neutralino is in the right ballpark (within an order of magnitude or so). This is the so-called *WIMP miracle*, where weakly interacting massive particle (WIMP) refers to any particle with properties similar to the neutralinos, whether associated with supersymmetry or not.

- Supersymmetry is an essential ingredient in realistic string theories, which allow a consistent quantum gravity and its unification with the other interactions. This does *not* require supersymmetry to survive down to the TeV scale, however: it could be broken anywhere below M_P.

If supersymmetry really exists at a low enough mass scale, it should be relatively easy to spot. The squarks

and gluinos carry color and could be produced in pairs
by QCD processes. They would then typically decay in
a series of stages (a *cascade*), such as $\bar{G} \to \bar{q}q\ell^+\ell^- +$
(unobserved) via

$$\bar{G} \to \bar{q}\bar{q}, \ \bar{q} \to q\bar{\chi}_2^0, \ \bar{\chi}_2^0 \to \ell^+\bar{\ell}^-, \ \bar{\ell}^- \to \ell^-\chi_1^0,$$
(6.1)

where $\bar{G}$, $\bar{q}$, $\bar{\ell}^\pm$, and χ^0 denote gluino, squark, charged
slepton, and neutralino, respectively. Pairs of such cascades
could lead to complicated final states, involving unobserved energy and momentum (assuming that χ_1^0 is the
stable LSP, which does not trigger the detector) as well as
multiple jets and/or leptons. No such events were spotted
at the Tevatron or the first run of the LHC. Will they be
seen at the second, higher-energy, run?

Unlike the SM, the Higgs mass M_H is constrained
in supersymmetry because there is no elementary quartic
coupling (see page 122), implying an *upper* bound
$M_H \lesssim 130\,\text{GeV}$. The observed $125\,\text{GeV}$ is consistent with
this bound, but that certainly does not prove super-
symmetry. In fact, most of the superpartner mass range
prefers smaller M_H. Supersymmetry also requires a second
Higgs doublet, so that there should be a second scalar,
a pseudoscalar, and a $\pm$ pair of charged Higgs particles,
perhaps in the several hundred GeV range.

Finally, supersymmetry has new and potentially problematic mechanisms associated with the new superpartners
for flavor changing processes and electric dipole moments.
This is also a concern for most other types of TeV-
scale physics (see the discussion beginning on page 157).
However, the nonobservation of BSM physics so far has

pushed the mass scales higher, aggravating the naturalness problem but somewhat alleviating the FCNC and EDM problems (and the tension with M_H for supersymmetry).

Many other types of (nonsupersymmetric) global, gauge, or discrete symmetries are possible. For example, *family symmetries* that relate the families and restrict the Higgs-Yukawa couplings are often postulated in an attempt to understand the fermion masses and mixings, especially those of the neutrinos. New symmetries could also explain small Dirac neutrino masses if they forbid the relevant Higgs-Yukawa interactions to leading order. Other symmetries are invoked in models of compositeness or strong coupling. Still others may emerge from the breaking of underlying high-scale unification theories, leading, e.g., to additional neutral or charged gauge bosons, perhaps at the TeV scale.

Compositeness and Strong Coupling

Perhaps the quarks and leptons are not truly elementary point-like particles, but are instead bound states (composites) of even smaller particles. After all, normal matter is made of of atoms and molecules, which themselves consist of electrons and nuclei. Nuclei in turn are bound states of nucleons, and nucleons are composed of quarks. Are there still more layers to this onion?

Perhaps there are, but any further levels of compositeness have hidden themselves very effectively. In atoms, molecules, and nuclei, the binding energies are small compared to the masses of the constitutents. Even the quarks in the proton and neutron have weak coupling when they are close together, and both their Higgs- and QCD-induced masses are comparable to or smaller than that of the nucleon.

If the quarks of leptons have further structure, it should manifest itself in the details of scattering processes like $e^+e^- \to e^+e^-$ or $e^+e^- \to q\bar{q}$ (Eichten et al. 1983), or in the production and decay of excited states. However, extensive searches for such effects at e^+e^- and hadron colliders have failed to observe any deviation from point-like behavior. Depending on the assumptions concerning the binding forces, these observations can be interpreted as lower limits of $\mathcal{O}(1-10 \text{ TeV})$ on the masses M of any constituents, roughly corresponding to quark or lepton radii being $\mathcal{O}(1/M) \lesssim 10^{-17} - 10^{-18}$ cm. The constituents would therefore have to be very much heavier than the bound states and have enormous binding energies! Any deviation from point-like would therefore have to be of an entirely different character from what we have encountered before. We will see an example in superstring theory.

TeV-scale bound state effects are more likely to be associated with spontaneous symmetry breaking than with the quarks and leptons, in part because the weak interaction scale $v \sim 246 \text{ GeV}$ is so much larger than most of the fermion masses. Some early dynamical schemes dispensed with the Higgs field altogether, instead breaking the electroweak symmetry by a *vacuum condensate* of $\bar{Q}Q$ pairs, i.e., $\langle 0 | \bar{Q}Q | 0 \rangle \neq 0$, where the *techniquarks* Q are heavier analogs of the quarks acted on by a new *technicolor* interaction.[3] Such alternatives had difficulties explaining

[3]This is similar to the breaking of the chiral extension of isospin in QCD.

the fermion masses, and in any case were essentially excluded by the discovery of the Higgs boson. Still viable, however, are *composite Higgs models*, in which strong TeV-scale dynamics does not directly break the electroweak symmetry. It instead generates the Higgs doublet as a bound state. New TeV-scale particles are generally expected in such theories, as well as small deviations in the properties of the Higgs boson. The relatively small value of M_H can be arranged if it is an approximate Nambu-Goldstone boson of the strong dynamics.

New Dimensions

The notion that there could more than three space dimensions was introduced in the 1920s by Theodor Kaluza and Oskar Klein. The motivation was to unify general relativity with electromagnetism by considering a fourth space dimension that is curled up and too small to easily perceive. For example, the gravitational metric G^{mn} ($m, n = 0, 1, 2, 3, 4$) in five space-time dimensions could include the ordinary metric $G^{\mu\nu}$ ($\mu, \nu = 0, 1, 2, 3$); the electromagnetic potential $A^\mu \equiv G^{\mu 4}$; and an additional scalar G^{44}. The Kaluza-Klein unification never worked out, but the idea of extra dimensions has reemerged in both top-down and bottom-up physics. They are an intrinsic ingredient in superstring theories, although there they are usually incredibly small, e.g., of $\mathcal{O}(M_P^{-1} \sim 10^{-33}\text{cm})$.

We already introduced extra dimensions as a possible solution to the Higgs-hierarchy problem in chapter 5 (page 144). Suppose, for example, that there are $3 + \delta$ space dimensions, with the extra ones occupying a finite volume $\propto L^\delta$, and that the fundamental scale associated with gravity is $M_F \ll M_P \sim 10^{19}$ GeV. To get a feeling for how this works, consider two masses $m_{1,2}$ separated by a distance r. For $r \gg L$, they feel the ordinary Newtonian gravitational force $\frac{m_1 m_2}{M_P^2 r^2}$. However, for $r \ll L$ the field lines spread out in the extra dimensions, yielding $\frac{m_1 m_2}{M_F^{2+\delta} r^{2+\delta}}$. Matching these expressions at $r = L$ implies

$$M_P^2 \sim M_F^{2+\delta} L^\delta. \qquad (6.2)$$

If the extra dimensions are *large*, i.e., small on everyday terms but large compared to the simplest expectation $L \sim M_F^{-1}$ from dimensional analysis, then one finds $M_F \ll M_P$. Since M_F would provide the cutoff for the Higgs mass-squared in (5.1), the Higgs-hierarchy problem would be reduced or solved for M_F in the TeV range (Arkani-Hamed et al. 1998), though at the expense of introducing a new puzzle as to why L is so large.

For $M_F \sim 10$ TeV, for example, one finds $L \sim 10^{12}$ cm, 10^{-3} cm, and 10^{-8} cm for $\delta = 1, 2$, and 3, respectively. $\delta = 1$ is clearly excluded, but larger δ could allow a small M_F. An obvious test of such theories is to search for the modification of the gravitational force law on distance scales smaller than L. Since gravity is so weak, it is difficult to probe sub-mm scales in the laboratory. Nevertheless, ultrasensitive modern versions of the Cavendish torsion balance experiment have ruled out any significant deviation on scales larger than around 10^{-3} cm (Adelberger et al. 2009), somewhat restricting M_F for $\delta = 2$. There are also Kaluza-Klein excitations of the graviton, associated with the quantized momentum in the extra dimensions, with apparent masses of $\mathcal{O}(1/L) \sim 10^{-17}$, 10^{-2}, and 10^3 eV for $\delta = 1, 2, 3$ and $M_F \sim 10\,\text{TeV}$. These are constrained by laboratory and astrophysical considerations (Olive et al. 2014), favoring somewhat larger M_F.

In more complicated geometries, there may be some large dimensions, and also some much smaller ones with size of $\mathcal{O}(M_F^{-1})$, which is around 10^{-18} cm for $M_F = 10\,\text{TeV}$. Some or all of the standard model particles might be able to propagate in these small dimensions, implying Kaluza-Klein excitations of $\mathcal{O}(M_F)$, which could be produced singly or only in pairs (depending on the details of the model) at the LHC. Other particles might be pinned to our three-dimensional boundary.[4] The fermion mass spectrum, including the possibilitiy of small Dirac neutrino masses, might be related to wave function overlaps in the extra dimensions.

In these large-dimensional scenarios, M_F is suppressed compared to M_P by a power law in $(L M_F)^{-1}$. An alternative, which leads to exponential suppression and thus avoids the need for large L, is that the extra-dimensional space is *warped*. That is, the metric and therefore the gravitational strength vary rapidly in the extra dimension, e.g., due to large energy sources at the boundaries. In the original version (Randall and Sundrum 1999), there is one extra dimension of length L, with all of the SM

---

[4]For technical reasons, most extradimensional theories are formulated with complicated topologies. A boundary to a dimension need not imply that one could "fall off the edge" or that there is another world on the other side. Rather, the "other side" might simply be some sort of mirror image.

particles fixed at one boundary, while the graviton wave function is peaked at the other. Later versions only fixed the Higgs at a boundary, with the fermions and gauge bosons allowed to propagate throughout the space. In both cases, the fundamental scale in the five-dimensional space is of $\mathcal{O}(M_P)$. However, the space warping modifies the physics at the boundary in such a way that

$$M_F \sim e^{-LM_P} M_P \qquad (6.3)$$

acts like a fundamental scale and cutoff for our apparent four-dimensional world, with $M_F \sim 10\,\text{TeV}$ for a very small $L = \mathcal{O}(35/M_P) \sim 10^{-31}$ cm. The weakness of gravity can be thought of as the result of the tiny overlap of the graviton wave function with our world.

An important consequence of warped dimensions is the existence of TeV-scale Kaluza-Klein excitations of the graviton, and of the fermions and gauge fields if they propagate in the extra dimension. These have been searched for at the LHC, e.g., by their decays into SM particles. These and other constraints are reviewed in Olive et al. 2014.

Warped dimensions may also be connected in various ways to electroweak symmetry breaking. *Higgsless models* accomplished the symmetry breaking by boundary conditions in the extra dimensions. These are ruled out by the observation of the Higgs boson, but the ideas may be applicable to the breaking of underlying unification symmetries. Another connection involves the composite Higgs models, briefly mentioned earlier, in which the Higgs fields are generated as bound states by strong dynamics at the

TeV scale. It turns out that some of these models are physically equivalent to weakly coupled extra-dimensional models, e.g., in which the Higgs is a component of a gauge field in five dimension. This is an example of a *duality*, that is, the physical equivalence of two apparently different descriptions of the same system.

This discussion hints at the enormous number of possibilities for extra dimensions, including their numbers, sizes, and topologies; which particles are free to move and which are stuck at specific points or boundaries; and whether they are flat or warped.

We briefly mention extra dimensions of a completely different character: supersymmetry can be formulated as field theory in a larger *superspace*, which includes the four ordinary dimensions as well as at least two additional complex coordinates that anticommute. Because of the anticommuting nature, Taylor series expansions in the new coordinates must truncate after a finite number of terms, i.e., one cannot move very far in the space.

(Almost) Hidden Worlds

It is a an intriguing possibility that there are one or more *hidden* sectors of nature to which we are blind, or nearly so. What I have in mind is less extreme than the parallel universes of science fiction or a variant on the multiverse idea. Rather, there could be new sets of fundamental particles that do not feel any of the SM interactions except gravity.[5] Conversely, they could have interactions amongst

---

[5] As far as we know, gravity couples to all forms of energy.

themselves that do not affect the SM particles. There could also be *quasi-hidden* sectors, which, in addition to gravity, communicate weakly with the ordinary particles by the exchange of a few *mediators*, which interact with both the SM and the hidden sector. In some cases, these are very heavy, while in others they may be light but have highly suppressed couplings to one sector or the other.[6]

The first major motivation for a hidden sector came from supersymmetry. We have seen that if supersymmetry exists at all, it must be broken. However, the explicit soft-breaking mass and cubic-scalar terms that maintain the advantages of supersymmetry are unlikely to emerge directly from an underlying superstring theory. They may, however, result from some kind of spontaneous breaking in a more complete theory. For technical and phenomenological reasons, the breaking is likely to occur in a hidden sector, perhaps associated with a vacuum condensate of pairs of the fermionic partners of some new hidden-sector gauge bosons. The breaking then generates soft terms in the supersymmetric SM, either by supergravity (i.e., by effects related to gravity by supersymmetry) or by other mediators.

Another possibility concerns the dark matter that constitutes most of the mass of the Universe, discussed starting on page 155. Although there are plausible dark matter suspects associated with supersymmetry (neutralinos) or the strong CP problem (axions), it is also possible that the

---

[6]For example, *dark gauge bosons* couple to the hidden sector. However, loop diagrams involving heavy particles could induce mixings with the ordinary gauge bosons, leading to small interactions of the dark bosons with our world, or of the ordinary gauge bosons with the hidden world.

dark matter lives in its own hidden (dark) sector. Such a dark sector could contain many particles and interactions, just as the SM sector does, with the stable one(s) being the dark matter.

Finally, hidden sectors could be random, i.e., not necessarily associated with any SM problem.

Hidden or quasi-hidden sectors may appear ad hoc. However, they can easily emerge in superstring or other underlying theories. By their very nature, however, they are difficult or impossible to detect. A truly hidden sector would have only gravitational effects, and could most likely never be directly detected in the laboratory. More encouraging are the various quasi-hidden sector models. If we are lucky enough, these could reveal themselves in laboratory and collider dark matter searches.

Unification

One of the most enticing ideas in physics is the possible unification of the fundamental interactions, so that they are seen to be different aspects of a simpler fundamental theory. Maxwell's unification of electricity and magnetism (section 2.5) was the first triumph. Einstein spent much of his later life attempting to construct a unified theory of electromagnetism and gravity, but the time was not ripe. We now understand that the strong and weak interactions are also essential players in any unification scheme. Furthermore, the success of the standard model, in which the strong, electromagnetic, and weak interactions are all described by gauge theories (and which partially unifies the weak interactions with electromagnetism), suggests that a promising starting point might be to join these three in a simpler grand unified theory (GUT). Grand unification does not incorporate quantum gravity, and (as in the SM) the Higgs-Yukawa interactions are rather ad hoc. Superstring theories go a step further and incorporate all of the known interactions in a single framework[7] that may or may not include a GUT.

Grand unification refers to a gauge theory in which $SU(3) \times SU(2) \times U(1)$ is part of a larger group with a single gauge-coupling constant, so that all of the SM gauge interactions are closely related as parts of the underlying theory. The simplest GUT is the Georgi-Glashow $SU(5)$ model (Georgi and Glashow 1974), which involves 24 gauge bosons. Twelve are the SM bosons (8 gluons; $W_{1,2,3}$; and B), while twelve more, known as the X, Y, and their antiparticles, are predicted. The X and Y transform as an $SU(2)$ doublet and carry electric charges 4/3 and 1/3, respectively. They are also triplets under $SU(3)$ and therefore come in three colors.

It is convenient to describe the $SU(5)$ properties of the L-chiral fermions and antifermions (rather than the L and R fermions), since they transform into each other. The properties of the R-chiral ones are related by a CP transformation. The fermions of each family fit into two $SU(5)$ representations, which are respectively 5 and 10 dimensional. The (ν_{eL}, e_L^-) and three colors of d_L^c are all

---

[7]An alternative approach, *loop quantum gravity*, focuses on the quantum properties of space-time and treats gravity separately from the other interactions.

related by $SU(5)$ in a 5* representation.[8] Here, d_L^c denotes a left-chiral anti-down quark, the CP-conjugate of d_R. Similarly, the u_L^c, (u_L, d_L), and e_L^+ transform as a 10. Schematically,

$$W^\pm \updownarrow \begin{pmatrix} \nu_L & \overset{X,\,Y}{\longleftrightarrow} & \\ & & d_L^c \\ e_L^- & & \\ & 5^* & \end{pmatrix} \begin{pmatrix} & \overset{X,\,Y}{\longleftrightarrow} u_L \overset{}{\longleftrightarrow} & \\ e_L^+ & & u_L^c \\ & d_L & \\ & 10 & \end{pmatrix} \updownarrow W^\pm,$$

(6.4)

where $SU(2)$ charged current transitions act vertically and the new transitions associated with X and Y act horizontally. [The $SU(3)$ color transformations act on the quark color indices, which are not displayed.] A right-chiral neutrino, ν_{eR}, can be added as an $SU(5)$ singlet, but is not required. The two heavier families have analogous transformations.

The fermions are treated in a more elegant way in the larger $SO(10)$ group of rotations in 10 dimensions, in which all of the fermions in each family, including the now required ν_{eR} (actually its CP-conjugate ν_{eL}^c), are joined in a single 16-dimensional representation. Even larger groups, which involve additional interactions and new fermions in each family, are also possible, and all of these theories can be extended to supersymmetric versions (e.g., Raby 2009; Langacker 2012; Olive et al. 2014).

---

[8]For $SU(n)$, $n > 2$, there is a difference between the n-dimensional representation, with matrices λ_i generalizing those in table 4.1, and its conjugate n^*, with matrices $-\lambda_i^*$. The CP-conjugates (e_R^+, ν_{eR}^c) and d_R transform as the 5.

Since the standard model interactions are observed to be so different in character, their underlying unification must be well hidden. This suggests that $SU(5)$ is spontaneously broken to the SM group at some large *unification scale* M_X, i.e., that the X and Y acquire the common mass M_X. This can occur if some components of a 24-dimensional Higgs representation acquire VEVs of $\mathcal{O}(M_X/g_5)$, where g_5 is the $SU(5)$ gauge coupling. At energy scales above M_X, the $SU(5)$ gauge symmetry is manifest: all of the SM and new gauge bosons and their interactions are related, and the $SU(3) \times SU(2) \times U(1)$ gauge couplings are given by $g_s = g = \sqrt{5/3}\, g' = g_5$, where the $\sqrt{5/3}$ ensures that the $SU(5)$ charges all have the same normalization. The 12 SM gauge bosons are not affected by the $SU(5)$ breaking: for scales lower than M_X but well above the electroweak scale they are effectively massless. The running $SU(3)$ $SU(2)$ and $U(1)$ gauge couplings all evolve at different rates, leading to the observed differences in their values at low energies. The breaking of $SU(2) \times U(1)$ to QED can be accomplished by a second stage of symmetry breaking at $\nu \sim 246\,\text{GeV}$ by the ordinary Higgs mechanism, where the Higgs doublet is assigned to an $SU(5)$ five-plet, along with a new colored scalar $\mathcal{H}$ with charge $-1/3$. The $\mathcal{H}$ must be superheavy [of $\mathcal{O}(M_X)$] to avoid rapid proton decay. Similar features hold for $SO(10)$ and similar GUTs.

An immediate consequence of this scenario is gauge unification, i.e., that the three SM running gauge couplings, determined experimentally at low energies and extrapolated theoretically to high-energy scales, should all meet at M_X (Georgi et al. 1974). This worked rather well

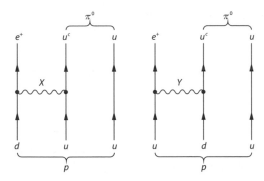

Figure 6.2. Typical diagrams for the proton decay process $p \to e^+\pi^0$ in the $SU(5)$ model, from Langacker 2010.

when $SU(5)$ was first proposed, with a very large predicted unification scale $M_X \sim 10^{14}$ GeV. It was less successful when one utilized later, more precise, measurements of the low-energy couplings, as shown in figure 6.1 on page 178. However, it came close enough to suggest that some modifications to the simplest theory, e.g., involving additional particles or a more complicated spectrum near M_X, might come to the rescue. In fact, the new particles in the supersymmetric extension of the standard model do exactly that, as can be seen in the lower plot in figure 6.1. In this case, $M_X \sim 3 \times 10^{16}$ GeV, which is (marginally) low enough compared to the Planck scale $M_P \sim 10^{19}$ GeV to justify the neglect of gravity.

The X and Y gauge bosons are superheavy. However, they can mediate unusual new *diquark* transitions between quarks and antiquarks, and also *leptoquark* transitions between quarks and antileptons. Taken together (figure 6.2), these can lead to proton decay processes such as

$p \to e^{+}\pi^{0}$, with a lifetime

$$\tau \sim \frac{M_{X,Y}^{4}}{\alpha_{5}^{2} m_{p}^{5}}, \qquad \alpha_{5} \equiv \frac{g_{5}^{2}}{4\pi}. \qquad (6.5)$$

They can also mediate decays like $n \to \bar{\nu}\pi^{0}$ for otherwise stable bound neutrons. The implications of (approximate) proton stability and limits on its lifetime were touched on page 158. The original $SU(5)$ model with $M_{X} \sim 10^{14}$ GeV suggested a lifetime $\sim 10^{29}$ yr and motivated several large proton decay searches, such as the SuperKamiokande water Cerenkov experiment and its predecessors. Proton decay was *not* observed at this level, with current lower limits on the lifetime of 10^{31}–10^{34} yr, depending on the decay mode. In the supersymmetric version of $SU(5)$ the lifetime due to X and Y exchange is much longer ($\sim 10^{38}$ yr) due to the larger unification scale, but there are new proton decay mechanisms associated with the fermionic superpartner of the $\mathcal{H}$, which lead to more rapid decays into modes like $\bar{\nu}K^{+}$. These have also not been observed, seriously challenging the simpler versions of supersymmetric grand unification.[9]

Despite its elegance and the success of gauge unification, GUTs have a number of difficulties. The nonobservation of proton decay and the omission of gravity have already been mentioned. We also glossed over two new theoretical problems. One is that the interactions between the Higgs 24-plet, which breaks $SU(5)$, and the 5, which leads to $SU(2) \times U(1)$ breaking, would typically result

---

[9] For a general review of proton decay, see, e.g., Nath and Fileviez Perez 2007.

in the two scales being close together. For them to be separated by 12 to 14 orders of magnitude requires a severe fine-tuning.[10] A second fine-tuning is needed to separate the Higgs doublet mass from that of its $SU(5)$ partner $\mathcal{H}$, which would otherwise lead to extremely rapid proton decay with a lifetime $\sim 10^{-11}$ s. This *doublet-triplet* problem can be avoided in some extensions of $SU(5)$ to higher dimensions.

Since quarks, leptons, and their antiparticles live together in $SU(5)$ multiplets, one might expect some relations between their masses. In the Georgi-Glashow model m_τ and m_b (or, more precisely, the Higgs-Yukawa couplings they are associated with, as in [4.49] and [4.50]) are predicted to be equal at M_X. These are running quantities just like the gauge couplings, and the observed low-energy values in table 3.2 are consistent with that prediction. However, the analogous predictions for the first two families fail badly, necessitating the complication of the model by additional very large Higgs representations or higher-dimensional operators. Similarly, large Majorana masses for the ν_R in $SO(10)$, needed for a neutrino seesaw (page 136), are usually obtained from a very large Higgs representation. Other fermion mass relations can be obtained by combining GUTs with additional family symmetries.[11]

---

[10]The supersymmetric version removes the large loop corrections to M_H^2 in (5.1) on page 142 but does not explain the origin of the weak scale in the first place.

[11]Family symmetries could possibly emerge directly from a larger GUT group, but this idea has not been very fruitful.

Grand unified theories predict the existence of super-heavy (mass $\sim M_X/\alpha_5$) *magnetic monopoles*, which are stable configurations of gauge and Higgs fields. These could have enormously overclosed the present Universe if they were initially produced in equilibrium density. However, a subsequent period of inflation (page 44) would have diluted their number density to essentially zero.[12] Another cosmological implication is for baryogenesis (page 152). The decays of the colored partner $\mathcal{H}$ of the Higgs and its antiparticle could have been CP-asymmetric,

$$\Gamma(\mathcal{H} \to q^c q^c) < \Gamma(\mathcal{H}^c \to qq),$$
$$\Gamma(\mathcal{H} \to q\ell) > \Gamma(\mathcal{H}^c \to q^c \ell^c), \tag{6.6}$$

apparently leading to an excess of quarks (Yoshimura 1978). This mechanism failed when it was realized that nonperturbative effects (page 153) would wipe out the asymmetry, but the basic idea was later resurrected in the leptogenesis mechanism (page 154).

The jury is still out on grand unification in its original form, so let us turn now to the even more ambitious superstring idea,[13] which may incorporate some or all of the aspects of GUTS, but also brings quantum gravity into the game.

Field theory is a remarkably successful framework for combining (special) relativity and quantum mechanics. However, field theory is plagued by infinities. These can

---

[12]This was one of Alan Guth's original motivations for proposing inflation (Guth 1981).

[13]Introductions include Zwiebach 2009; Dine 2015.

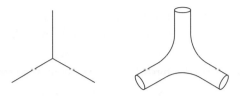

Figure 6.3. Left: The worldlines of three particles meet at a point in field theory. Right: The merging of the worldsheets (i.e., tubes) of three closed strings.

be sidestepped in renormalizable theories, where the divergences do not appear in expressions for physical quantities when expressed in terms of measured parameters (page 13), and in any case the divergences appearing in intermediate steps become finite if the troublesome integrals are cut off at a new physics scale Λ_{NP} (page 142). However, straightforward attempts to quantize general relativity are horribly nonrenormalizable, and it is not obvious that there is any new physics scale higher than the Planck scale to cut off the divergences.

The infinities in field theory can be traced to the assumption that the particles are point-like. Consider, for example, the three-particle interaction vertex in a field theory, illustrated by the Feynman diagram in figure 6.3. The lines can be interpreted in position space as the worldlines of the three particles. The interaction occurs when these meet at a point in space-time, and the amplitude involves the integration over the positions of each of these interaction points. The integration includes singular points at which the interactions overlap, leading to the bad high-energy behavior and divergences.

In string theory, the particles are no longer point-like. Rather, they are tiny one-dimensional extended objects known as *strings*. These may be open or closed, and have characteristic length $l_s = 1/M_s$, where the *string scale* M_s is usually assumed to be of the order of the Planck scale, $M_P \sim 10^{19}$ GeV ($1/M_p \sim 10^{-33}$ cm). The strings can vibrate in a variety of normal modes. Most have masses of $\mathcal{O}(M_s)$, but there are some "massless modes," which can be interpreted as fundamental particles. They appear pointlike when probed on distance scales large compared to l_s, and can acquire small masses by the Higgs or similar mechanisms.

Strings sweep out two-dimensional *worldsheets*[14] as they propagate in space-time. Interactions correspond to the splitting or fusing of these worldsheets, as shown in figure 6.3. Since such string junctions do not occur at isolated space-time points, they do not lead to any singularities, and are therefore finite (at least to each order in string perturbation theory). From the point of view of the approximate corresponding field theory, the strings provide a natural cutoff $\Lambda_{NP} \sim M_s$. There is one important caveat, however: there are subtle quantum consistency conditions (anomaly cancellations) that are satisfied only (for theories involving fermions) if there are ten dimensions of space-time. In addition to time and our ordinary three large and flat (or nearly flat) space dimensions, there must be six additional ones, presumably curled up (*compactified*) in a small compact manifold, similar to those introduced in some

---

[14]The worldsheets of closed strings are actually tubes.

bottom-up scenarios (see the discussion starting on page 183). Furthermore, the existence of chiral fermions requires that these dimensions have complicated topologies.

Unlike field theory, there is qualitatively only one type of particle (the string) and one interaction, the string junction in figure 6.3, suggesting that all of the fundamental particles and interactions are unified and described by simple geometrical considerations. There are no free dimensionless parameters, so one might think that string theory leads to unique predictions for the low-energy theory, such as for the gauge group and other symmetries, the particle content, the values of the gauge and Yukawa couplings, and for the ratios of the string scale M_s to other dimensional parameters, such as M_P (only such dimensionless ratios are observable). In fact, there was a time (in the 1980s) when some physicists thought it would provide a finite, parameter-free, unique *theory of everything*.

We now know that this view was overoptimistic and oversimplified. For one thing, there are five known supersymmetric string theories (i.e., superstring theories). Five is small compared to the infinite number of possible field theories, but superstring theory is nevertheless not unique. The theories differ in the allowed bosonic and fermionic excitations and in whether the strings are open or closed. All contain a massless spin-2 closed string mode, which can be identified with the graviton. Superstring theory therefore automatically includes quantum gravity, which is finite perturbatively! Other massless modes have other spins and quantum numbers, and may correspond to fermions, gauge bosons, etc. In one (*heterotic*) superstring

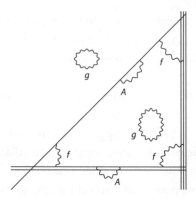

Figure 6.4. Massless modes in an intersecting brane construction. Stacks of D-branes occupy one dimension in a two-dimensional slice of the extra dimensions. Gauge bosons (*A*) are described by strings terminating on a single stack, chiral fermions and their superpartners (*f*) are localized at intersections, and gravitons (*g*) are freely propagating closed strings.

theory, these are closed. They may propagate freely in all ten dimensions (like the graviton), or may be trapped at isolated singularities in the extra dimensions. There are also various massive modes, including vibrational excitations, Kaluza-Klein excitations, and winding modes (in which the strings wrap around the extra dimensions, such as around a donut).

String theories may also contain nonperturbative higher-dimensional objects known as *D branes*. For example, the *Type IIA* superstring contains membrane-like objects that span six space dimensions: the three large ones and three of the six compactified dimensions. Open strings can terminate on these D branes, as illustrated in figure 6.4.

Strings beginning and ending on a stack of n parallel D branes may be the gauge bosons of an $SU(n) \times U(1)$ gauge theory. They are free to propagate in the ordinary dimensions and to slide along the stack in the extra ones. In one (*intersecting brane*) construction chiral fermions and their superpartners are localized at the intersections of two stacks and are charged under both gauge groups. They can move freely only in the ordinary space-time dimensions.

The five superstrings theories are not entirely independent: some are physically equivalent to each other, to different limits of the same theory, or to entirely different theories, in the sense that they have the same spectrum, interactions, and other observables. Such equivalences between theories that appear to be completely different are known as dualities. For example, one theory characterized by an extra-compact dimension of size L may be dual to another with a dimension α'/L, where $\alpha' \propto M_s^{-2}$ is known as the string tension. There is a correspondence between the states of the two, such as between Kaluza-Klein and winding modes. Other dualities involve replacing the string coupling g_s (related to M_s/M_P) by $1/g_s$, i.e., so that the weak coupling limit of one theory is related to the strong coupling limit of another. Moreover, the five theories are all thought to be special limiting cases of a more general M ("mystery") theory, which also includes 11-dimensional supergravity, but at present this underlying theory is not well understood.

Even more troubling than the existence of five types of superstring is that for each there are an enormous number of possible solutions to the basic equations, corresponding to a large landscape of (mainly metastable) vacua, and the physics depends on which vacuum one is in. Furthermore, there is no known selection principle to single one or the other out. Many of the differences are associated with the sizes, shapes, and topological properties of the extra dimensions. The vacua can also differ in the numbers, dimensions, and locations of the D branes; background fluxes (analogs of the electromagnetic fields); and other nonperturbative effects. Many of the parameters, e.g., involving the sizes of the extra dimensions, are actually given by the expectation values of scalar modes (*moduli*), somewhat like the spontaneous breaking of $SU(2) \times U(1)$ that determines the electroweak scale and related quantities (as described in section 4.6). In string theory, however, there is an added complication that these expectation values are not fixed at the perturbative level, implying a continuum of possible vacua, at least until nonperturbative effects are included. For example, the string coupling g_s is determined by the expectation value of e^ϕ, where ϕ is a modulus known as the *dilaton*. The gravitational coupling (i.e., the ratio M_s/M_P) and gauge couplings are determined by g_s and (in some cases) by the volumes of the extra dimensions and of the D branes [cf. equation (6.2)]. Yukawa couplings also depend on what is going on in the extra dimensions. In the Type IIA constructions in figure 6.4, for example, Yukawa couplings between particles at the corners of a triangle of area A in the extra dimensions are proportional to $\exp(-A)$.

It has been suggested that there may be $\mathcal{O}(10^{600})$ vacua, although the number could be much larger or smaller. The skeptic could well argue that the theory of *everything* has morphed into a theory of *anything*. I do not think that this is entirely fair, but this is a good time to step back and comment on some of the pros and cons of string theory:

- String theory provides a unification of quantum mechanics with gravity! Furthermore, gravity is unified with other interactions, and the divergences that have plagued field theory for many decades have disappeared. It is the only known game in town for accomplishing all of these feats, and the importance should not be underestimated.

- However, the particle physics community has not yet come to grips with the implications of the huge landscape of string vacua, for which we have no known uniqueness or selection principle to determine which is the one chosen by nature. The vast majority of these are not realistic, e.g., because the gauge symmetry or particle content is smaller than the SM, or because they contain more or fewer than three large space dimensions.[15] However, many are consistent with everything we know.

- The string landscape might, however, be the ideal home for the environmental or anthropic considerations discussed in sections 5.2 and 5.3. It is possible that there is no unique vacuum. Rather, many or all of the vacua might have been produced at various times and places in a multiverse through some kind of eternal inflation, but life could only evolve in the tiny fraction for which the conditions are "just right."

---

[15]Only three conventional space dimensions allow stable planetary orbits, with possible implications for environmental selection (Linde 2015).

For example, the value of the cosmological constant in typical string vacua is probably $M_s^4/M_P^2 \sim 10^{120}$ times larger than the observed one, but if the values in 10^{600} vacua are randomly distributed, then some would "miraculously" have the small value needed for complex structures to form (see page 162). One must be cautious, however: it is not certain how the values of the cosmological constant are distributed, and most of those that have been studied have negative dark energy, rather than the positive value that is observed. Also, it is not certain that the vacua are really metastable. Fortunately, questions like this will probably be resolved by further theoretical work.

- The biggest problem is that it is difficult to make concrete predictions to confirm or falsify string theory. This is partly because the string scale is usually enormously higher than scales that will ever be directly probed in the laboratory. Even more daunting is the multiplicity of vacua. Finally, although string theory is conceptually simple, the technical details involved in constructing a consistent and realistic theory are horrendous. For these reasons, many people have assumed that string theory cannot be tested even in principle and have dismissed it as "unscientific." I personally reject that point of view, partly because of the more general comments starting on page 172, and also because of its great conceptual success with quantum gravity and unification. Moreover, I now turn to experimental possibilities that could significantly strengthen or weaken the case for string theory, even if it were not rigorously established or falsified.

Barring an unanticipated technical breakthrough, it is unlikely that we will ever be able to locate the "true vacuum" of nature in the string landscape, either by somehow zeroing in on it or by brute-force exploration of all of the possibilities. Nevertheless, string theory is actually more restrictive than field theory. A large number of the vacua that have been explored predict new physics beyond the SM, which tends to fall into a more limited number of possibilities than the corresponding explorations in field theory. Discovery of such "string preferred" effects would certainly argue for string theory, while observation of "string disfavored" effects would strongly argue against.

String theories can be constructed with or without supersymmetry, but those involving fermions (and with perturbatively stable vacua) are supersymmetric. Supersymmetry can therefore be considered a fairly generic prediction, although the supersymmetry breaking scale could be far above the TeV scale. Many string vacua also share some, but not necessarily all, of the predictions of grand unification. For example, many lead to proton decay at some level. There are sometimes predictions concerning gauge unification and Higgs-Yukawa couplings, but these often differ in detail from those of simple GUTs. This is because the predictions depend on the string origin of the massless spectrum, and because the grand unification symmetry (if present) may be broken in the compactification of the extra dimensions rather than by the Higgs mechanism.

Massive string excitations can have spin larger than two.[16] We will never be able to produce these directly (unless M_s is very low), but they may have been produced during an inflationary period in the early Universe, and their decays may have left a tiny but in principle detectable imprint on astrophysical observables (Arkani-Hamed and Maldacena 2015)!

Supersymmetry, GUT-like predictions, and higher spins are examples of new physics that would be suggestive of string theory. On the other hand, very large gauge groups and, more important, very large representations are examples of "string disfavored" effects. For example, grand unified theories frequently invoke very large Higgs representations, such as a 126-dimensional representation of $SO(10)$ that is utilized in many seesaw models for neutrino mass. However, such large representations are incredibly rare in the string landscape (if they exist at all). The reason is that the basic gauge degrees of freedom in string constructions are the low-dimensional (fundamental) representations and those that can easily be constructed from them. For example, in the intersecting brane constructions, each string carries one fundamental charge at each end. The associated state is either bifundamental (under two groups), adjoint (one charge and one anticharge), or a symmetric or antisymmetric combination of two fundamentals, just as the total spin of two spin-1/2 particles can only add up to 0 or 1. Therefore, indirect evidence for large representations in grand unification or

---

[16] One can view a string theory as being equivalent to a field theory involving an infinite tower of higher-spin states, with interactions related in such a way as to cancel all divergences.

direct evidence for such states at lower energy would constitute strong evidence against string theory. There are additional (more subtle) consistency relations involving the low-energy spectrum that are more stringent in string theory than in field theory.

There are also possible "remnant" signatures of many string vacua. By this, I mean some new particles, interactions, or effects that slip through into the low-energy theory in specific compactifications, essentially by accident, and that do not necessarily solve any standard model problem or satisfy minimality. These are extremely common in classes of string vacua that have been examined. Among the most frequently encounted are additional $U(1)$ gauge bosons, extended Higgs sectors involving $SU(2)$ singlets and/or more than the two doublets of the MSSM, and new "exotic" fermions and their superpartners with nonstandard $SU(2) \times U(1)$ assignments. For example, there could be new charge $-1/3$ quarks in which both the left- and right-chiral states are $SU(2)$ singlets (i.e., heavier versions of the d, s, and b but with no charge $2/3$ partner). These sometimes decay by mixing with ordinary quarks, but sometimes they involve unusual new interactions, allowing decays into a quark and lepton or into two antiquarks. Other common implications of string vacua include very light and weakly coupled spin-0 axions, similar to those suggested for the strong CP problem; flavor-changing or family nonuniversal effects, e.g., because the three fermion families sometimes do not have the same string origin; and hidden or almost hidden sectors, such as might be associated with supersymmetry breaking or dark matter.

String theory has suggested a novel physical mechanism for small masses. One often finds that certain types of masses or interactions are forbidden at the perturbative level by symmetries imposed by the string vacuum, but are generated nonperturbatively. These are exponentially suppressed compared with M_s, with the exponent depending on the volumes of certain D brane configurations. For example, exponentially small Dirac neutrino masses may be generated, in contrast to the power-law suppression of the seesaw model. Similar (but less suppressed) effects could lead directly to the higher-dimensional Majorana mass operator in (4.69). There have also been several realizations of the neutrino seesaw model, involving both exponential and power-law suppressions, but the details are usually more complicated than in bottom-up constructions.

These are examples of the most commonly encountered types of new physics in known string vacua. Unfortunately, though suggestive they are neither necessary nor sufficient for establishing string theory. Much work has also been done on finding vacua that incorporate the MSSM but nothing else. Such efforts could possibly shed light on likely patterns of the fermion masses and mixings, for example, though no particularly compelling results have yet emerged. On the other hand, new strong dynamics at the TeV scale could also lead to some of the same types of remnants as the string constructions. If we are so lucky as to observe interesting TeV-scale physics, it will be nontrivial but perhaps not hopeless to discern whether it is more likely associated with string theory or with strong dynamics.[17]

---

[17]I am here implicitly assuming that our sector of physics has no strong dynamics between the electroweak and string scales. String theory could also

I briefly mention the possibility that the string scale M_s is very much lower than M_P (e.g., Berenstein 2014), which would be allowed as long as l_s is smaller than the experimental limit on quark and lepton sizes of around 10^{-17}–10^{-18} cm (page 182), i.e., $M_s \gtrsim$ 1–10 TeV, which is low enough that string excitations and other effects might be directly observable in the laboratory! However, similar to the bottom-up discussion of large dimensions starting on page 183, values of $M_s \ll M_P$ would require new fine-tunings involving the sizes of the extra dimensions.

Whether or not string theory turns out to be the correct description of gravity and the other interactions, the mathematics that it has generated, and in particular the understanding of dual interpretations of the same phenomena, has provided important tools for other branches of physics. These include quark-gluon plasmas (which probe QCD at high temperatures and density) and the description of exotic states in condensed matter (Sachdev 2013).

Here Be Dragons

As radical as string theory might appear, with its extra dimensions, branes, and possible connection to the multiverse, it is really a rather conservative extension of field theory. In its simplest form, it maintains the conventional notions of space and time, at least for the $3 + 1$ ordinary dimensions, and of quantum mechanics and special relativity, while gauge bosons and interactions result in a

---

lead to strong dynamics, but in that case we might never be able to see through that layer of nature to the underlying string theory.

fairly straightforward way from string vibrational modes
and string junctions. Perhaps a more dramatic break with
traditional concepts is needed, such as the introduction of
nonlocal hidden variables into quantum mechanics (Adler
2014) or the violation of causality.

It is possible that some of the things we take for
granted are really *emergent*, that is, that they are not
fundamental but appear only as approximations to the
underlying physics. A familiar example is heat, which is
not a fundamental fluid (as was once believed) but is rather
associated with the random motions of large numbers
of atoms and molecules. Similarly, space may be emergent (e.g., Seiberg 2006). We have seen a hint of this in the
string duality mapping a theory with a compact dimension
L onto an equivalent theory with dimension α'/L, i.e.,
the spatial size is somehow not fundamental. Even the
number of dimensions is not sacred: there is a powerful
duality between certain four-dimensional field theories
and five-dimensional gravitational theories, known as the
Anti-de Sitter/conformal field theory (AdS/CFT) correspondence[18] (Maldacena 1999). I finally mention that in string
theory nothing can probe distances shorter than l_s, even
in principle (somewhat reminiscent of the uncertainty
principle). Perhaps space is just an approximation valid at
large distances.

Interactions may also be emergent. For example, there
are examples of gauge theories that are dual to other
nongauge theories, while the AdS/CFT correspondence

---

[18]More generally, *holography* refers to the concept that the information in a
spatial volume may be encoded on its surface, in analogy with optical holography.

suggests that even gravity may be emergent. Let us leave these heady speculations, however, and turn to the experiments.

6.2 The Tests

There are several approaches to searching for new phenomena in high-energy physics. One is to go to higher energies, using energetic particles produced at new accelerators or from astrophysical sources such as cosmic rays. This direction, the *energy frontier*, allows broad exploration and tests of the standard model and beyond, and especially seeks to directly produce very massive new particles. In recent years, the Tevatron and LHC have led this effort, carrying out detailed studies of QCD in the weak coupling regime; of the properties of the heaviest SM particles, such as the W, Z, and t; and discovering the Higgs boson. Another approach is the *intensity frontier*, which tests the SM with as much precision as possible. The goals include the accurate determination of the SM parameters; the search for small deviations from its predictions; and seeking evidence for processes that are rare or forbidden in the SM. The intensity frontier encompasses such diverse domains as low-energy QED; weak decays and properties of the muon; precision Z-pole experiments at LEP and the SLC; studies of flavor physics at B factories; and studies of neutrino mass and mixing. Finally, the *cosmic frontier* refers to the areas of overlap between astrophysics/cosmology and particle physics. This includes cosmological, astrophysical, and laboratory observations relevant to the dark matter, dark energy, and neutrinos; the direct observation of neutrinos produced in the Sun and constraints from the Sun or stars on the existence other light particles; and searches for the annihilation products of exotic heavy objects such as magnetic monopoles. In addition to the experimental and observational opportunities, there have been advances in our theoretical tools. These include powerful computer programs to automate the often tedious calculations needed to predict the consequences of the SM and BSM physics, and mathematical breakthroughs that enormously simplify the calculation of scattering amplitudes that would have been hopelessly complicated using traditional Feynman diagram techniques (e.g., Arkani-Hamed and Trnka 2014).

These approaches have always been complementary and overlapping. Advances in technology have expanded the opportunities on each front in recent decades and for the forseeable future.

The Laboratory

The LHC (described on page 119) will be the flagship of the energy frontier program for the immediate future. The collider restarted in 2015 in a second phase that should continue through 2022 (including a planned shutdown near the middle). It is expected to have an energy of 13–14 TeV, nearly twice that of the first run, and should eventually reach ten times the *luminosity* (a measure of the relative flux of the two beams).

The higher energy is especially important for a pp (or $\bar{p}p$) collider because the most interesting processes involve the collisions of constituent quarks, antiquarks, and gluons, and each of these typically carries only a fraction of the energy of the parent proton (e.g., Barger and Phillips 1997). While it might be kinematically possible for a pp collision at 8 TeV to produce a 4 TeV particle, for example, this would rarely occur because the probability of the constituents carrying that much energy is small. At 13 TeV, however, the constituents would need a smaller fraction of the proton energy, and the rate would be much higher. The event rate also increases linearly with the luminosity. The higher energy and luminosity make it easier to separate the interesting signals from more mundane obscuring background processes that can lead to similar final states.

This next phase of the LHC should roughly double the reach of the first run for supersymmetry, strong coupling, remnants, and dark matter particles, e.g., from $\gtrsim 1$ TeV for many colored particles to twice that. Any new physics that was just out of reach in the first run should show up quickly. Another order of magnitude increase in the luminosity of the LHC is anticipated in a third (14 TeV) phase, from around 2026 to 2035. This will further extend the reach, as well as allow more detailed studies of any new phenomena previously discovered.

If new particles are observed at the LHC, they would likely only be the lower-mass tip of the iceberg of a new sector; if they are not observed, the LHC reach would still not have sufficient energy to definitively answer the question of whether nature chooses naturalness (section 5.3). Fortunately, superconducting magnet technology has progressed far enough for us to seriously consider even higher-energy

pp colliders. There is a proposal at CERN to construct an extremely large new circular collider. This would involve the construction of a new 80–100 km circumference tunnel that would run around Geneva, partly underneath the lake. Following a first e^+e^- phase, the tunnel would house an 80–100 TeV pp collider, beginning operation around 2040 or 2050. There is a somewhat similar proposal in China for a circular collider to be built either some 300 km east of Beijing or near Hong Kong. These machines would have a reach for colored particles up to around 30 TeV. In addition to probing most of the likely regime of naturalness and the associated BSM physics, such facilities would be similarly sensitive to remnants and to dark matter particles, and could explore the possibility of electroweak baryogenesis (page 154) through detailed study of the Higgs self-interactions.

There are exciting parallel prospects on the intensity frontier. There is a proposal for a 31-km-long linear e^+e^- collider (ILC) in Japan, with an initial CM energy of 350–500 GeV and an upgrade to 1 TeV possible. The pp proposals for CERN and China each involve an e^+e^- collider in the same tunnel as a first stage. These would be of somewhat lower energy than the linear collider (up to 350 GeV at CERN) but would have higher luminosity. Any of these would allow more precise measurements of the couplings of the Higgs boson, which are very sensitive to extended Higgs sectors, alternative compositeness models, and new particles entering higher-order corrections. They would also probe new physics in other ways. For example, new heavy $U(1)$ gauge bosons could be detected and studied in $e^+e^- \to \mu^+\mu^-$ through their interference with diagrams involving the γ and Z, even for masses considerably exceeding the e^+e^- CM energy. This would be especially powerful if the new gauge boson were also produced directly at the LHC, with the two types of observations giving complementary information on the couplings. These facilities would also have the option of running at lower energies, i.e., at the (91 GeV) Z pole, where they could produce 10^2–10^5 more Z's than were observed at LEP. This would allow a repeat of the whole Z-pole program described in section 4.7, but with a precision improved by as much as a factor of 100 and correspondingly increased sensitivies to all kinds of new physics.[19]

I also mention the CLIC proposal for an $\sim$3 TeV e^+e^- linear collider to run at CERN after the LHC. CLIC would have a novel acceleration mechanism: there would be two beams in each direction, with one beam deriving extra energy from the electromagnetic fields produced by the other high-current *drive beam*. CLIC would mainly be sensitive to new BSM physics. There is also talk of a future $\mu^+\mu^-$ collider, which could be of much higher energy than e^+e^- because the muons radiate less energy. However, plans are less developed than for the other projects.

All of these proposals are technically feasible, but it is difficult to give definite predictions for their prospects

---

[19]This would, however, require a major theoretical effort to calculate the SM expectations to higher order.

or timescales. They are all quintessential examples of "big science": they are very large; complex; expensive to build (typically billions of dollars) and run (enormous amounts of electrical power are needed); involve large international collaborations; and require the cooperation of many governments, which is not so easy in today's world. The projects are also to some extent in competition with each other for financial and human resources. Presumably, however, the future e^+e^- programs other than CLIC would run in the 2020s and 2030s, in parallel with the LHC, while future pp colliders would take over in the succeeding decades. Such timescales are not particularly long compared with historical developments in the field, but are nevertheless starting to become comparable to a person's professional lifetime. One must be patient.

It would probably not be feasible to ever go to much higher energies than those just mentioned by the same technologies. There are, however, ideas being developed that could eventually shrink the physical size of particle accelerators greatly. In particular, particles could be accelerated rapidly by plasma waves, which could be created by a laser pulse or by injecting bunches of electrons or protons. It is too early to speculate on the prospects for such technology.

The intensity frontier also encompasses neutrino physics, flavor physics, and rare processes. A very active neutrino program is under way or proposed. This includes long- and medium-baseline neutrino oscillation experiments in the United States, Japan, China, and Korea that should measure the amount of leptonic CP violation, determine the mass ordering (hierarchy), and yield more precise mixing angles (see the discussion beginning on page 128). The most ambitious of these is the Deep Underground Neutrino Experiment (DUNE) proposal to send a high-intensity neutrino beam from Fermilab to the Sanford Underground Research Facility in Lead, South Dakota (the site of the original Davis Solar neutrino experiment), some 1300 km distant.

There is also an active effort searching for the neutrinoless double beta decay process $nn \to ppe^-e^-$ ($\beta\beta_{0\nu}$). This would violate lepton number by two units, and could occur by the exchange of a virtual Majorana neutrino. The existence or not of $\beta\beta_{0\nu}$ would establish whether the neutrinos are Majorana or Dirac. However, observation of such a rare process requires significant quantities (e.g., $10-10^4$) kg of appropriate isotopes, which must be ultrapure and well-shielded in underground labs[20] to avoid radioactive and cosmic ray–induced backgrounds. Planned experiments would unfortunately only be sensitive to the inverted neutrino hierarchy (page 133) or to the case in which the masses are all large compared to $\Delta m_\odot^2$. Additionally, refined cosmological studies and a new tritium β decay experiment (KATRIN in Germany) should push the sensitivity to the absolute mass scale down to the few tenths of an eV level suggested by the mass-squared differences from oscillation studies. A number of reactor or radioactive source experiments should settle whether

[20] A number of types of experiments, including $\beta\beta_{0\nu}$, Solar and atmospheric neutrinos, proton decay, and direct dark matter detection, must be conducted deep underground to shield from cosmic rays. These are usually located in mines or automobile tunnels. Multi-experiment underground facilities include Gran Sasso in Italy, Kamioka in Japan, the Sanford Lab in South Dakota, Soudan in Minnesota, and SNOLAB in Ontario.

eV-scale sterile neutrinos exist (page 136). Together, this program of experiments should nail down the properties of the neutrino mass and mixing, which is essential for testing the specific models proposed and possibly shedding light on the validity of the uniqueness paradigm (section 5.3). They will also be sensitive to BSM physics, such as new interactions, neutrino decay,[21] significant magnetic moments, or sterile neutrino mixing.

The neutrino program will continue to be of astrophysical interest as well. Large underwater or under-ice experiments, such as the IceCube experiment (an array of photomultipliers frozen in a km^3 of ice at the South Pole; e.g., Gaisser and Halzen 2014), probe high-energy neutrinos from astrophysical or exotic particle physics sources, but also have some sensitivity to oscillations. Other experiments study Solar neutrinos in detail, perhaps even those from rare secondary reactions, or search for a burst of neutrinos from a core-collapse supernovae in our galaxy or a background of supernova neutrinos from distant galaxies. It is even possible that future tritium β decay experiments could detect the *relic neutrinos* left over from neutrino freezeout (page 46) when the Universe was around one second old. (These are analogous to the CMB photons, which froze out much later.)

Some of the detectors for the long-baseline neutrino experiments will do double duty searching for proton decay. The DUNE experiment will eventually employ a 40 kT liquid argon detector nearly a mile underground,

---

[21]Heavier neutrinos can in principle decay radiatively, e.g., $\nu_2 \to \nu_1 \gamma$, in the SM though loop diagrams, but for the allowed mass and mixing parameters the lifetime is far longer than the age of the Universe.

while the Hyper-Kamiokande experiment will employ a 1000 kT water Cerenkov detector, some 20 times larger than its predecessor SuperKamiokande. These should be able to extend the sensitivity to proton decay by an order of magnitude, further into the ranges suggested by grand unification and some superstring vacua.

Other experimental programs are also sensitive to new physics. There are active programs at the LHC, at Fermilab, at the J-PARC and KEK facilities in Japan, and elsewhere studying heavy quark properties, searching for flavor-changing processes such as $\mu^- N \to e^- N$ and $K^+ \to \pi^+ \nu \bar{\nu}$, repeating the muon anomalous magnetic moment experiment (page 64) with more precision, and searching for electric dipole moments.

Cosmology and Astrophysics

We are also in a golden age for astrophysics and cosmology, but I will mention only a few things most relevant to particle physics. The dark energy (page 155) and dark matter (page 156) will be studied in considerable detail by the Dark Energy Survey (DES) and by the Large Synoptic Survey Telescope (LSST), which are respectively 4 and 8 m telescopes high in the Chilean Andes. They will survey Type I supernovae, study subtle correlations of the distributions of galaxies, produce detailed three-dimensional maps of the entire sky, and study dark matter–induced strong and weak *gravitational lensing*. These observations will collectively map the time evolution of the dark matter distribution and of the expansion of the Universe, hopefully distinguishing between alternative possibilities for the dark energy. The nature of the dark matter is also constrained by continuing studies of the size distributions of galaxies and their density profiles. Primordial black holes can be searched for by gravitational effects such as microlensing.

There is also a multipronged program to identify dark matter through nongravitational effects. A number of *direct detection* experiments are searching for the (weak) scattering of dark matter particles from ordinary matter in the laboratory. These experiments have excluded some of the likely parameter space for WIMPs and for lighter candidates such as those associated with a dark sector (page 189), and will ultimately be limited in sensitivity by backgrounds from Solar and atmospheric neutrinos. *Indirect detection* refers to γ rays or other particles (especially positrons or antiprotons) produced by the annihilation (or slow decay) of dark matter particles in astronomical sites such as the Sun or galactic center. There have been several hints of indirect detection, but the challenge is to separate the signals from ordinary astrophysical backgrounds. For example, there is a significant excess of γ rays from the galactic center. This may turn out to be due to the annihilations of dark matter particles in the 60 GeV range, but it is also possible that they are from pulsars or other compact sources. Finally, dark matter particles can be searched for via unobserved energy produced in association with ordinary particles at the LHC. For example, a $\bar{q}q$ pair could annihilate to produce an unobserved WIMP particle with a gluon radiated from one of the initial quarks, leading to a single *monojet* of hadrons.

Axions are extremely light and weakly coupled spin-0 particles often predicted in string theories or as a possible solution to the strong CP problem; in some cases they could constitute some or all of the dark matter. They would be extremely difficult to detect, but nevertheless there is an active program searching for axions (whether or not relevant to dark matter), mainly involving their electromagnetic coupling (Graham et al. 2015). (They couple to $\vec{E} \cdot \vec{B}$, as in [5.2].) A number of experiments search for cosmological axions or axions produced in the Sun by their conversion in an applied magnetic field, exciting a high-Q microwave cavity. Other possibilities involve time-dependent induced EDMs, or "light shining through a wall." In the latter case, photons are converted to weakly coupled axions in an applied $\vec{B}$ field, and then reconverted to photons by another $\vec{B}$ field on the other size of an opaque barrier.

The remarkably detailed studies of the CMB described in section 3.3, combined with other astrophysical probes, have provided a wealth of information concerning dark matter, dark energy, new forms of radiation, and neutrino masses. A number of further investigations are under way, many focusing on photon polarization. Certain patterns would be evidence for primordial gravity waves. If these were seen, they would provide observational information about a period of the Universe far earlier than anything else available. In particular, they would strongly support the notion that there was an initial period of inflation that smoothed out the Universe, provided the seeds for galaxies through quantum fluctuations, and perhaps even populated the multiverse.

I finally mention the first direct observation of gravitational waves (from merging black holes) at the Laser Interferometer Gravitational-Wave Observatory (LIGO) (Abbott et al. 2016), heralding the field of gravitational wave astronomy. Although primarily of interest for studying violent astrophysical events and testing general relativity, the lack of dispersion in the LIGO event already implies a new upper limit of $\sim 1.2 \times 10^{-22}$ eV on the graviton mass, and future observations could possibly detect signals from effects such as phase transitions in the early Universe.

7

EPILOGUE: THE DREAM

Humankind's thirst for knowledge has led (in the realm of physics) to an understanding of the laws of motion and gravity, electromagnetism, kinetic theory, relativity, quantum mechanics, and atomic and nuclear physics; has contributed greatly to chemistry, biology, and materials science; and has led (usually as a by-product) to a myriad of practical applications. Parallel and frequently interconnected progress in astronomy has allowed an understanding of the nature of the Solar System and stars, the existence of galaxies, the expansion of the Universe, and of its origin in the big bang.

As recounted in some detail in this volume, recent decades have witnessed the development and testing of the standard model of elementary particles and their interactions, which synthesized all that had been learned previously into an elegant but complicated theory. The standard model is mathematically consistent and correctly describes most aspects of nature down to distances of around 10^{-16} cm, an achievement that was almost undreamed of in the 1960s. The standard cosmological model was developed in roughly the same period; it successfully describes the large-scale structure of the Universe and traces its evolution in detail back to the time when it was ~ 1 s old. The two standard models are closely intertwined because the early Universe consisted of a hot dense plasma of elementary particles.

These successes are an enormous achievement, and together they answer most of the questions that I had so many years ago as a graduate student. Nevertheless, both theories are incomplete. The problems of the particle physics model were detailed in chapter 5. Similarly, we think we have at least the broad outline of what must have happened in the Universe prior to 1 s, but this is largely inferred from the particle physics—there are no direct observational tests prior to the big bang nucleosynthesis era. Many of the details, such as the origin of the baryon asymmetry, are tied to to the shortcomings of the SM. We still do not know the details of the first inflationary instant that may have smoothed out the Universe, or even whether inflation really occurred. And what lies beyond the 14 billion light-year horizon? Is our Universe perhaps part of a multiverse? What caused the big bang in the first place, and was there something even earlier?

We have seen in chapter 6 that there are many promising ideas about what might underlie the standard model. It is my dream that we will some day develop and substantiate a new "Standard Model of Nature," incorporating physics in an elegant unified framework up to the Planck scale and describing the Universe on the largest scales and back to the big bang. This will not be an easy task. It will require substantial experimental and observational advances from the bottom-up, combined with major theoretical progress from the top-down. It may not be possible if nature does not cooperate, and it may not be feasible if we do not have sufficient ingenuity, dedication, and luck. However, we are intelligent beings endowed with curiosity about our place in the scheme of things. We should try.

POSTSCRIPT: RUN 2

When the original manuscript for this book was completed in November, 2015 the LHC had recently started its Run 2, at higher energy but (initially) lower luminosity than the first run, but no results had yet been announced. Everyone eagerly awaited the first news, to be presented at a seminar at CERN by the ATLAS and CMS collaborations on December 15 and broadcast worldwide over the internet, with many hoping that evidence for supersymmetry or one of the other popular ideas for saving naturalness would be announced. In the event, there was no hint of any of the anticipated kinds of new physics, although in fairness the limited statistics available at that time offered relatively little improvement in sensitivity over Run 1. But the ATLAS and CMS people had a surprise for us. Each group had measured the number of diphoton ($\gamma\gamma$) events as a function of their total mass and found an excess around 750 GeV. While the statistical significance was not compelling for either experiment, the fact that they seemed to be seeing the same thing at the same mass strongly suggested that the signal was real. The $\gamma\gamma$ signal was reminiscent of the Higgs boson discovery, but the details did not fit well for the particle to be one of the heavy Higgs bosons predicted in supersymmetric models. Perhaps a new and unexpected particle far heavier than anything seen before had been discovered!

Over the next several months the theoretical community worked feverishly to try to explain the new particle. It turned out that it could easily be accomodated as an "add on" to the standard model, to the MSSM, or to a number of other theories, that is, a particle that was not required but was just there, such as a top-down remnant. For example, it could have been a 750 GeV spin-0 particle that coupled to photons and to gluons (for the production process) via loop effects involving new heavy non-chiral quarks. Further, more detailed, studies might reveal an entirely new sector of Nature.

Alas, it was not to be. The next significant announcement of LHC results was at the biennial International Conference on High Energy Physics, held in Chicago in August, 2016. By then, the experiments had greatly increased their statistics, and the $\gamma\gamma$ excesses had disappeared. They were just a statistical fluke after all. The 750 GeV story will be a tiny footnote to the history of particle physics, but I relate it to give a bit of the flavor of doing science. One must be patient and painstaking if one hopes to tease out Nature's secrets.

To add insult to injury, the LHC experiments had seen no indication of any other new physics, with the sensitivity increased significantly compared to Run 1. At about the same time, the LUX experiment, located in the underground Sanford lab in South Dakota, announced the results of the most sensitive search to date for WIMP dark matter: the data was consistent with background, and

LUX was able to set new stringent limits on the interaction strength of any WIMPs over a wide mass range.

So where does that leave us? On the positive side, the lack of any direct indication of new physics can be viewed as a triumph for the standard model. It works more precisely and over a larger range of mass and distance scales than we had thought. However, all of the shortcomings of the standard model remain to be explained. As detailed in Chapter 6 the planned and proposed programs at the LHC and elsewhere will continue for many years, and will extend the sensitivity much further. We may still find the new physics that was postulated to maintain naturalness, and we may still directly establish the nature of WIMP or axion dark matter. Or, Nature may surprise us with something totally unexpected. Or, perhaps there is nothing new at the multi-TeV scale, supporting the landscape/multiverse ideas. Any of these possibilities are fascinating. I am as eager as ever to find out.

GLOSSARY

α: Fine structure constant $= e^2/4\pi \sim 1/137.04$, where $-e$ is the electric charge of the electron.

Anomalous magnetic moment (a): The correction from higher-order interactions to a particle's magnetic g factor. For a fermion, $a = (g - 2)/2$.

Anomaly: Symmetries of a classical field theory are sometimes broken by quantum corrections known as anomalies. Uncancelled anomalies in a gauge theory violate renormalizability.

Antiparticle: An antiparticle has the same mass but the opposite values for quantum numbers such as electric charge, color, strangeness, and lepton number as the corresponding particle. Antimatter refers generically to antiparticles.

Asymptotic freedom: The property of QCD that α_s becomes smaller at short distance.

Axion: A possible very light and weakly coupled spin-0 particle associated with some solutions to the strong CP problem and also by some superstring theories. Some or all of the dark matter could be in the form of axions.

Baryogenesis: The dynamical generation of an excess of baryons with repect to antibaryons (the baryon or matter asymmetry). Necessary ingredients (the Sakharov conditions) are baryon number (B) violation, CP violation, and either departure from equilibrium of the B-violation processes or CPT violation. A number of possible mechanisms have been proposed.

Baryon: A half-integer spin hadron, usually consisting of three quarks. Baryons carry baryon number $B = 1$, while the antibaryons have $B = -1$.

BBN: Big bang nucleosynthesis. The synthesis of the light elements ^4He, D, ^3He, and ^7Li in the first few minutes after the big bang, when the Universe had cooled sufficiently, to $T \sim 0.1$ MeV, for nuclei to be bound.

Beta (β) decay: Radioactive decay in which $n \to pe^-\bar{\nu}_e$ or $p \to ne^+\nu_e$, usually in a nucleus.

Big bang: The (now generally accepted) theory that the Universe began around 14 billion years ago, when it was extremely hot and dense, and has been expanding and cooling ever since. The term was introduced by Fred Hoyle to contrast it with the now-discredited steady state theory.

BNL: Brookhaven National Laboratory. A research institute located on Long Island, New York.

Boson: A particle with a symmetric multiparticle wave function. Bosons in three dimensions have integer spin.

Bottom-up: Models directly motivated by experimental or phenomenological considerations, usually involving new physics at the TeV scale.

Branching ratio: The fraction of times that a particle decays into a particular final state.

Brane: A nonperturbative membrane-like object in string theory, spanning $D \geq 0$ space dimensions.

BSM: Beyond the standard model. Possible new physics that incorporates and extends the standard model.

c: Speed of light in vacuum $\sim 3.0 \times 10^{10}$ cm/s ($c = 1$ in particle units).

CERN: Conseil Européen pour la Recherche Nucléaire. International high-energy physics lab located on the Swiss-French border near Geneva. Site of the LHC, LEP, the SPS $p\bar{p}$ collider, and other experiments.

Charge conjugation (C): The replacement of a particle by its antiparticle. C invariance is violated maximally by the WCC.

Chirality: Left- and right-chirality refer to the projections $(1 \mp \gamma^5)/2$ on a Dirac spinor. These are mapped onto each other by space reflection, but transform differently under a chiral symmetry. Chirality coincides with helicity for a relativistic spin-1/2 particle.

CKM: Cabibbo-Kobayashi-Maskawa matrix. The 3×3 unitary mixing matrix that describes flavor change and CP violation in the quark part of the WCC. The 2×2 submatrix for the first two families is the Cabibbo rotation, involving the Cabibbo angle $\sin\theta_c \sim 0.23$.

CM: Center of mass (actually, center of momentum). Reference frame in which the two particles in a collision have equal and opposite momenta.

CMB: Cosmic microwave background radiation. Photons (currently at 2.7 K) left over from the early big bang, when electrons and nuclei combined to form atoms (recombination).

Color: The analog of electric charge for QCD. Gluons are emitted or absorbed by colored particles. Each flavor of quark can carry either red (r), green (g), or blue (b) color, with the names inspired by (but not related to) optical colors.

Compositeness: The possibility that (apparently) fundamental particles are actually bound states.

Confinement: The aspect of QCD that prevents quarks and gluons from existing as isolated particles.

Cosmological constant (Λ): A constant describing the energy of space (i.e., the vacuum), given by $\Lambda = 8\pi G_N \rho_{vac}$, where ρ_{vac} is the vacuum energy density. A positive cosmological constant by itself implies an exponentially expanding (de Sitter) universe.

Cosmology: The study of the characteristics and history of the Universe as a whole.

Coupling (constant): A number (or function of renormalization scale) describing the strength of an interaction vertex, such as the amplitude to emit or absorb a boson.

CP: The product of charge conjugation and space reflection. For example, a left-chiral electron is mapped onto a right-chiral positron. CP violation occurs at a level of 10^{-3} in the weak interactions.

CPT: The product of C, P, and T transformations. A local Lorentz-invariant unitarity field theory must be invariant under CPT (Streater and Wightman 2000).

Dark energy: A type of energy that does not change in density as the Universe expands, most likely due to energy stored in a scalar field. It may be constant in time (cosmological constant) or slowly varying (quintessence). From observations such as the acceleration of the expansion of the Universe and of the CMB, approximately 70% of the energy density of the present Universe is dark energy.

Dark matter: Matter with little or no interaction with light or ordinary matter. Its existence is inferred from its gravitational effects, such as on the dynamics of galaxies and clusters, gravitational lensing, and in the CMB. Approximately 25% of the energy density in the present Universe is dark matter, while 5% is ordinary matter. Its detailed composition is unknown.

DESY: Deutsches Elektronen-Synchrotron. Laboratory in Hamburg, Germany. Site of several accelerators, including the PETRA (e^+e^-) and HERA (ep) colliders.

Dirac mass: A fermion mass term that conserves particle number.

DIS: Deep inelastic scattering. Processes such as $e^- N \to e^- X$ or $\nu_\mu N \to \mu^- X$ at high momentum transfer from the leptons to the hadrons. The final hadrons X are summed over.

Discrete symmetry: A symmetry such as charge conjugation invariance that does not depend on a continuous parameter.

Domain wall: A field configuration at the boundary between regions with different vacua.

Duality: The equivalence (in terms of the physical predictions) of two apparently different theories. Often one theory involves strong coupling, while the other is weakly coupled and more tractable.

EDM: Electric dipole moment. A separation between positive and negative electric charge. EDMs for the elementary fermions violate P, CP, and T invariance. In the SM, the EDMs generated by the ordinary weak interactions are extremely small, but they could be larger due to BSM physics or the strong CP mechanism.

Eightfold way: An approximate $SU(3)$ global flavor symmetry of the strong interactions, which extends isospin to include strange particles. It is broken at the 25% level by quark mass differerences.

Electroweak scale (v): $v \equiv \sqrt{2}\langle 0|\phi^0|0\rangle \sim 246$ GeV describes the SSB of $SU(2) \times U(1)$ and sets the scale for the $W^\pm$ and Z masses.

Electroweak theory: The $SU(2) \times U(1)$ gauge theory of the weak and electromagnetic interactions, which incorporates QED and the Fermi theory.

Emergent: An apparent phenomenon (such as heat) that is not part of the fundamental theory, but instead arises from an underlying complexity.

Environmental selection (anthropic principle): The possibility that a parameter or other feature of a theory is not determined by any uniqueness principle, but rather because it allows for a complex and long-lived universe. This becomes plausible if there is a landscape of possible vacua and a means (such as eternal inflation) of sampling them.

Eternal inflation: A type of inflation in which new inflating regions are continually spun off. These may have different laws of physics, leading to a multiverse.

eV: Electron volt. The energy acquired by an electron accelerated through a one volt potential $\sim 1.8 \times 10^{-33}$ g c^2. Related units are 1 keV (kilo-eV) $\equiv 10^3$ eV, 1 MeV (Mega-eV) $\equiv 10^6$ eV, 1 GeV (Giga-eV) $\equiv 10^9$ eV, and 1 TeV (Tera-eV) $\equiv 10^{12}$ eV.

EWBG: Electroweak baryogenesis. The possible generation of the baryon asymmetry during the phase transition that led to spontaneous $SU(2) \times U(1)$ breaking.

Extra dimension: A space-time dimension in addition to the known four. New space dimensions are usually curled up (compact), and may be very small (e.g., size $= \mathcal{O}[M_P^{-1} \sim 10^{-33}$ cm$]$), large ($\lesssim 10^{-3}$ cm), warped (strongly curved by gravity), etc.

Family: The fundamental fermions (u, d; ν_e, e^-) or heavier copies with identical interactions. The other known families are (c, s; ν_μ, μ^-) and (t, b; ν_τ, τ^-). Family symmetries relate the families in an attempt to understand the flavor problem.

FCNC: Flavor-changing neutral current. A flavor-changing transition that preserves electric charge, such as $d \to s$ or $e^- \to \mu^-$. These do not occur at lowest order in the SM model because of the GIM mechanism (for the Z) and the minimal Higgs structure, but are a stringent constraint on some BSM theories.

Femtobarn (fb): Cross-section unit. $1\,\text{fb} = 10^{-39}\,\text{cm}^2 = 10^{-3}\,\text{pb}$ (picobarn) $= 10^{-15}$ barn.

Fermi (fm): 10^{-13} cm, the approximate size of a proton or neutron.

Fermi constant (G_F): A parameter describing the strength of the weak interactions at low energy. $G_F \equiv \sqrt{2}g^2/8M_W^2 \sim 1.2 \times 10^{-5}\,\text{GeV}^{-2}$, where g is the $SU(2)$ coupling constant.

Fermilab: Fermi National Accelerator Laboratory, near Chicago. Site of the Tevatron and various neutrino and other experiments.

Fermion: A particle with an antisymmetric multiparticle wave function that therefore obeys the Pauli exclusion principle. Fermions in three dimensions have half-integer spin.

Fermi theory: A theory of the weak interactions based on four fermions interacting at a point. It is nonrenormalizable but nevertheless an excellent first approximation to low energy decays and scattering.

Feynman diagram: A graphical representation of the terms in perturbation theory, with vertex factors corresponding to interactions and external (internal) lines corresponding to real (virtual) particles.

Field: A function of space and time typically associated with a force or particle. A free quantum field (which occurs in perturbation theory) is an operator that can act on a state to produce a new state with one more or one fewer particle.

Flat, closed, open universe: See Friedmann equation.

Flavor: A label for the quark and lepton mass eigenstates. The different flavors may differ in electric charge and in other quantum numbers such as strangeness. Flavor symmetries are approximate symmetries between different flavors, such as isospin or the eightfold way. The flavor problem refers to our lack of understanding of the fermion families, masses, and mixings.

Friedmann equation: $\left(\frac{\dot{R}}{R}\right)^2 = \frac{8}{3}\pi\rho G_N + \frac{\Lambda}{3} - \kappa c^2$, the equation for the scale parameter R in a homogeneous isotropic (Robertson-Walker) universe. ρ is the energy density in particles and radiation ($\rho \propto R^{-3}$ for nonrelativistic particles, while $\rho \propto R^{-4}$ for radiation, including relativistic particles). Λ, which is independent of R, is the cosmological constant (vacuum energy). κ is the curvature constant, which can be taken to be ± 1 or 0 for appropriate units. For $\Lambda = 0$, a flat universe ($\kappa = 0$) has a perfect balance between the expansion $(\dot{R}/R)^2$ and the energy density. It has no spatial curvature, and will expand forever, with $\dot{R}/R \rightarrow 0$ asymptotically. An open universe ($\kappa = -1$) will expand forever, while a closed universe ($\kappa = +1$) will eventually reach a maximum size and then collapse.

Gauge covariant derivative: A modified derivative involving a gauge field, such as $\partial^\mu + ig A^\mu$, used in constructing a gauge invariant theory.

Gauge (local) invariance: A symmetry or invariance that depends on the location in space and time. Gauge transformations are sometimes referred to as redundancies in the description because they act on unobservable quantities such as vector and scalar potentials in QED. Gauge invariant interactions are mediated by spin-1 gauge bosons.

Gauge unification: The possibility that the (properly normalized) running $SU(3) \times SU(2) \times U(1)$ gauge couplings all meet at some large unification scale M_X, above which symmetry-breaking effects can be ignored.

Gell-Mann matrices: The eight 3×3 matrices λ_i that describe the transformation of $SU(3)$ triplets, generalizing the $SU(2)$ Pauli matrices.

GIM: Glashow-Iliopoulos-Maiani mechanism. The absence of FCNC $d \leftrightarrow s$ transitions mediated by the Z at lowest order or of enhanced second-order effects involving two Ws, because of the existence of the c quark, which allows the s_L to transform as an $SU(2)$ doublet.

Global symmetry: A symmetry that is independent of space and time.

Gluon (G): The eight colored gauge bosons of QCD.

Gravitational or Newton constant (G_N): The constant describing the strength of the gravitational force. $G_N \sim 6.67 \times 10^{-8} \, \mathrm{cm}^3 \, \mathrm{g}^{-1} \, \mathrm{s}^{-2} = 6.71 \times 10^{-39} \, \mathrm{GeV}^{-2}$.

Gravitational lensing: The deflection of light, e.g., from a distant star or galaxy, as it passes through the gravitational field of the Sun, a galaxy cluster, or some other concentration of mass. Lensing is useful for mapping the distribution of dark matter. Strong lensing can produce rings, arcs, or multiple images, while the more subtle weak lensing distorts a single image. Small objects (e.g., planets) passing in front of a background star can temporarily amplify its light (microlensing).

Graviton (g): The hypothetical spin-2 particle that mediates quantum gravity.

Group: A set of objects (usually symmetry transformations in physics applications) with an associative multiplication law, an identity element, and a unique inverse. It is referred to as abelian if it is also commutative, and nonabelian otherwise.

GUT: Grand unified theory. A gauge theory in which the SM is embedded into a group such as $SU(5)$ with a single gauge coupling constant.

Hadron: A colorless strongly interacting particle, such as the proton or pion. The conventional hadrons are either bound states of three quarks (baryons) or of a quark and an antiquark (mesons).

$\hbar$: Reduced Planck constant $= h/2\pi \sim 6.6 \times 10^{-22}$ MeV-s ($\hbar = 1$ in particle units), which sets the scale for quantization. For example, the quantum of energy of an electromagnetic wave of frequency $\omega/2\pi$ is $\hbar\omega$.

Helicity: Projection of the spin of a particle onto the direction of its momentum. For a spin-1/2 particle, helicity $= +\frac{1}{2}$ or $-\frac{1}{2}$ is referred to as right-helicity or left-helicity, respectively (or as right- or left-handed). For a spin-1 particle, helicity ± 1 correspond to right or left-circular polarization, respectively. In addition to these transverse modes, a massive spin-1 particle can have longitudinal polarization (helicity $= 0$).

Hidden sector: A possible new sector of particles and interactions that are only weakly coupled to the SM.

Higgs boson (H): The 125 GeV spin-0 particle associated with the background Higgs field that generates mass for the $W^\pm$, Z, and chiral fermions (the Higgs mechanism).

Higgs-hierarachy problem: The puzzling fact that the Higgs mass (and electroweak scale) are many orders of magnitude smaller than the Planck (gravity) scale.

Higgs-Yukawa (Higgs) interaction: The interaction between fermions and the Higgs boson.

Higher-dimensional operator: A nonrenormalizable operator that is multiplied by inverse powers of a large mass scale, presumably associated with BSM physics.

Hubble parameter (H): $H \equiv \dot{R}/R$, where R is the scale parameter, describes the expansion rate of the Universe. The current value is $H_0 \sim 67$ km/s-Mpc, where 1 Mpc $\equiv 10^6$ pc and a parsec (pc) is $\sim 3.1 \times 10^{18}$ cm ~ 3.3 light-years.

Hyperon: A baryon with nonzero strangeness, but no c, b, or t quarks.

Inflation: A possible brief period of exponentially rapid expansion that smoothed and flattened the Universe, followed by a reheating to some large temperature T_{RH}.

Infrared slavery: The property of QCD that α_s becomes large at large distance, leading to confinement and the binding of quarks into hadrons.

Internal symmetry: A symmetry involving the intrinsic properties of particles, such as their phases or particle type. As opposed to space-time symmetries.

Isospin: An approximate $SU(2)$ global flavor symmetry of the strong interactions, relating, e.g., the proton and neutron. It is broken at the $\sim 1\%$ level by quark masses and electromagnetism.

IVB: Intermediate vector boson. The massive charged bosons $W^\pm$, especially in the ad hoc (nongauge) extension of the Fermi theory.

Jet: A cluster of hadrons produced by a high-momentum quark or gluon.

Kaluza-Klein particles: Excitations of ordinary particles with higher quantized momenta in extra dimensions.

Kaons ($K^\pm$, K^0, $\bar{K}^0$): The lightest mesons carrying strangeness ± 1. They have mass ~ 490 MeV, spin-0, and consist of $q\bar{s}$ or $s\bar{q}$, where $q = u$ or d.

Lagrangian (density): A function of the dynamical variables of a classical or quantum theory from which the equations of motion can be derived. In field theory, the Lagrangian density $\mathcal{L}$ typically includes terms for the kinetic energy and mass of a particle as well as interaction terms corresponding to vertices in Feynman diagrams.

Landscape: The collection of vacua in a theory with a large number of metastable or stable minimum-energy configurations. Superstring theory and some field theories may have landscapes.

LEP: Large Electron-Positron Collider. The 27 km circumference e^+e^- collider at CERN that operated from 1989 to 2000, with CM energy from 91 to 209 GeV.

Leptogenesis: A possible mechanism for baryogenesis, in which a lepton asymmetry is first created by the decays of very heavy Majorana neutrinos and then partially converted to a baryon asymmetry through nonperturbative effects.

Lepton: A spin-1/2 particle with lepton number $L = +1$, including the e^-, μ^-, and τ^- and their associated neutrinos. Their antiparticles have $L = -1$. The leptons and antileptons do not experience the strong interaction.

LHC: Large Hadron Collider. The 27 km circumference pp collider that began operations in 2010 at CERN, with a design CM energy of 14 TeV.

Little hierarchy problem: The fine-tuning required if there is new physics much heavier than the electroweak scale, but still far below the Planck scale.

LSP: Lightest supersymmetric partner. In many versions of supersymmetry, it is absolutely stable.

Magnetic monopole: An isolated magnetic charge. Monopoles can emerge in grand unified and similar theories as topologically stable configurations of gauge and Higgs fields.

Majorana mass: A fermion mass term that violates particle number (i.e., lepton number for neutrinos) by two units.

Mass eigenstate: A particle with definite mass, analogous to energy eigenstates in quantum mechanics.

Meson: An integer spin hadron, usually consisting of a quark and an antiquark.

Minimality: The assumption that a theory should be as simple as possible.

Moduli: Scalar modes in string theory whose expectation values determine aspects of the theory, such as coupling constants and the sizes of extra dimensions.

MSSM: The minimal supersymmetric extension of the standard model.

Multiplet: A set of quantum states that are related by a symmetry.

Multiverse: A postulated superset of our observed Universe, consisting of a large or infinite number of regions, each with different parameters or laws of physics.

Muon (μ): A charged fermion that is identical to the electron except for its mass, $m_\mu \sim 207 m_e$.

Nambu-Goldstone boson: The massless rolling mode associated with a spontaneously broken global symmetry. In a gauge theory, it disappears (is "eaten") and reemerges as the longitudinal polarization state of a massive gauge boson.

Naturalness: The assumption that the qualitative aspects of physics should not depend on a fine-tuning of parameters.

Neutrino: A nearly massless neutral fermion that feels only the weak interactions and gravity.

Neutrino oscillation: The oscillation of one flavor of neutrino into another due to the mismatch between weak and mass eigenstates.

Neutrinoless double beta decay ($\beta\beta_{0\nu}$): The process $nn \rightarrow ppe^-e^-$ in a nucleus. This would violate lepton number by two units, and could occur via $nn \rightarrow ppe^-e^-\nu\nu$ if the (virtual) neutrinos can annihilate through a Majorana mass term.

Noether theorem: The theorem that any continuous symmetry leads to a conserved current and charge.

Nucleon: A proton or neutron.

Nucleosynthesis: The production of nuclei heavier than the proton in the big bang, stars, core-collapse supernovae, or cosmic ray intereactions.

Parity (P): Refers either to a space reflection or to its eigenvalue ± 1. Parity invariance is violated maximally by the WCC.

Particle units: Convenient units used in particle physics, in which $\hbar = c = 1$. All physical quantities then have dimensions of energy to some power. For example, velocity (v) is dimensionless; energy (E), momentum (p), and mass (m) have units of energy; and distance (x) and time (t) have units of 1/energy. Ordinary units can be restored by multiplying a quantity by appropriate powers of $c \sim 3.0 \times 10^{10}$ cm/s, $\hbar \sim 6.6 \times 10^{-22}$ MeV-s, and $\hbar c \sim 197$ MeV-fm, and using 1 kg $\sim 5.6 \times 10^{29}$ MeV/c^2.

Photon (γ): The massless spin-1 particle associated with electromagnetism.

Pions ($\pi^\pm, \pi^0$): The lightest mesons ($m_\pi \sim 140$ MeV), which mediate the long-range part of the nuclear force. They have spin-0 and consist of $q_1\bar{q}_2$, where $q_{1,2}$ are u or d.

Planck mass or scale (M_P): The energy scale at which quantum gravitational effects are important. $M_P \equiv G_N^{-1/2} \sim 2.2 \times 10^{-8}$ kg $\sim 1.22 \times 10^{19}$ GeV in particle units. The related Planck length (time) is $1/M_P \sim 1.6 \times 10^{-33}$ cm (5.3×10^{-44} s).

PMNS: Pontecorvo-Maki-Nakagawa-Sakata matrix. The leptonic analog of the CKM matrix.

Point-like: A particle with no internal size or structure.

Polarization asymmetry: The difference in cross sections (divided by the sum) as the polarization of a beam of particles is reversed.

Positron (e^+): The antiparticle to the electron.

Propagator: The factor in a Feynman diagram corresponding to an internal line. It is proportional to $1/(q^2 - m^2)$ for a particle of mass m and virtual momentum q and may have additional spin factors.

Proton decay: The possible decay of the proton, e.g., into $e^+ \pi^0$. Proton decay is predicted in many grand unified and other theories. The experimental lower limit on the lifetime is $\sim 10^{31}$ yr.

QCD: Quantum chromodynamics. The $SU(3)$ gauge theory of the strong interactions of quarks and gluons.

QED: Quantum electrodynamics. The relativistic quantum theory of electromagnetism.

Quark: A spin-1/2 point-like particle carrying color. There are six known flavors, u, d, c, s, t, and b. Hadrons are color-neutral bound states of quarks and antiquarks.

Range: The distance between two particles for which an interaction is significant. For example, the strength is proportional to

$\exp(-mr)$ for interactions mediated by a particle of mass m, and the range is $1/m$.

Reheating: The conversion of vacuum energy into thermal energy following inflation.

Renormalization: The expression of observables in terms of the measured values of quantities such as mass and charge rather than the original (bare) parameters in the Lagrangian. A renormalizable theory involves only a finite number of such quantities.

Representation: A group representation is a concrete realization by matrices or operators of the multiplication law. $n \times n$-dimensional matrix representations are often used in physics to describe how n states or operators related by a symmetry are transformed into each other.

Running: The property that coupling "constants" actually depend on the energy (or renormalization scale) because of higher-order corrections.

Scalar: A spin-0 particle or its associated field, or a quantity that does not change under rotations. In some contexts, the term also implies that it is unchanged under space reflection, as opposed to a pseudoscalar, which does change sign. Scalar can also refer to a quantity that is invariant under isospin or other internal symmetries.

Scalar potential: A function of the spin-0 fields containing mass terms and nongauge interactions.

Scale factor (R): A length parameter whose time variation describes the expansion or contraction of the Universe.

Seesaw model: A model in which small Majorana masses for ordinary doublet neutrinos are generated by mixing with heavy singlet Majorana neutrinos.

SLAC: Stanford Linear (National) Accelerator Center in Palo Alto, California. Site of the two-mile linear e^- accelerator and several e^+e^- colliders.

SLC: Stanford Linear Collider. An e^+e^- collider that that ran at the Z pole (i.e., CM energy $= M_Z$) from around 1989 to 1998, with a polarized e^- beam starting in 1992. Electrons and positrons were accelerated in the SLAC two-mile linear accelerator and then diverted to collide head on.

SM: Standard model. The combination of quantum chromodynamics and the $SU(2) \times U(1)$ electroweak theory.

Soft supersymmetry breaking: Breaking by explicit mass terms and cubic scalar interactions, which do not spoil the supersymmetry relations between dimensionless couplings, such as those that cure the Higgs-hierarchy problem.

Space reflection: The transformation of a system into its mirror image (times a rotation). Classically, $\vec{x} \to -\vec{x}$, $\vec{p} \to -\vec{p}$, and $\vec{J} \to +\vec{J}$.

Spin-statistics theorem: The theorem in relativistic field theory that integer (half-integer) spin particles must be bosons (fermions).

SSB: Spontaneous symmetry breaking. A symmetry of the equations of motion that is not respected in the lowest energy state, such as the alignment of electron spins in a ferromagnetic domain. As opposed to explicit breaking of a symmetry by a small term in the equations of motion.

Standard cosmological model: The description of our Universe as being approximately flat, homogeneous, and isotropic, with the present energy density dominated by dark energy (70%), dark matter (25%), and ordinary matter (5%).

Strangeness (S): A flavor quantum number that is conserved in the strong and electromagnetic but not the weak interactions. The s quark carries $S = -1$.

String (superstring): A possible description of particles as vibrational modes of tiny open or closed one-dimensional extended objects. String theories incorporate quantum gravity.

String scale (M_s): The typical energy of massive modes of string vibrations, related to the string size by $M_s = l_s^{-1}$.

Strong coupling: A theory or regime in which the couplings are too large to allow a perturbative treatment, often leading to bound states.

Strong CP problem: The possibility that the neutron could receive a large electric dipole moment from subtle QCD effects.

Strong interaction: The strong short-range interaction between hadrons, responsible for nuclear binding and energy, or the underlying QCD interaction of quarks and gluons.

SU(n): The group of $n \times n$–dimensional unitary matrices with determinant 1. Rotations in three dimensions are described by $SU(2)$. The global flavor isospin and eightfold way symmetries are described by $SU(2)$ and $SU(3)$, respectively. The weak and strong interactions involve the gauged $SU(2)$ and $SU(3)$ groups. The simplest grand unified theory is based on $SU(5)$.

SuperKamiokande: A very large (50 kiloton) water Cerenkov detector located in a mine in western Japan, utilized for neutrino oscillation studies and proton decay searches. A larger future facility, Hyper-Kamiokande, is planned.

Supernova: An exploding star, caused either by matter falling onto the surface of a white dwarf from a companion object (Type I), or by the collapse of the core of a very massive star

(Type II). The former are useful for determining distance, while the latter produce heavy elements.

Superpartner: In supersymmetry, each particle is related to a superpartner differing in spin by 1/2 unit, with closely related interactions and (if supersymmetry is unbroken) the same mass. The spin-0 partners of the quarks and leptons are known as the squarks and sleptons, the spin-1/2 partners of the gluon is the gluino, and the spin-1/2 partners of the charged (neutral) electroweak Higgs and gauge bosons are the charginos (neutralinos).

Supersymmetry: A symmetry between fermions and bosons. A gauge (local) version is known as supergravity.

Symmetry (invariance): A transformation that leaves the properties or equations of motion of a system unchanged.

Tau (τ): A charged lepton heavier than the electron or muon, $m_\tau \sim 17 m_\mu$.

Tevatron: The 6.9 km circumference $\bar{p}\,p$ collider that operated at Fermilab from 1983 to 2011, with a final CM energy of $\sim 2\,\text{TeV}$.

Time reversal (T): The transformation of a system onto the time-reversed one. For a classical trajectory, $\vec{x}(t) \to \vec{x}(-t)$, $\vec{p}(t) \to -\vec{p}(-t)$, and $\vec{J}(t) \to -\vec{J}(-t)$.

Top-down: BSM physics that is motivated by fundamental considerations such as unification of the interactions, usually involving a very high energy scale such as M_P.

U(1): The group of phase factors. Applications include rotations in two dimensions, the global symmetries associated with baryon and lepton number, and the gauge symmetry of QED or the SM.

Unification: The combination of two or more seemingly different interactions as aspects of a more fundamental one, such as electric and magnetic interactions into electromagnetism.

V−A: Vector minus axial current. The form for the WCC involving the maximal amount of space-reflection and charge-conjugation violation.

Vector: A spin-1 particle or field, or a quantity such as $\vec{x}$ or $\vec{p}$ that acts like a classical vector under rotations. In some contexts, it is implied that the quantity also changes sign under space reflection, as opposed to an axial vector (e.g., angular momentum $\vec{J} = \vec{x} \times \vec{p}$), which does not change sign. Vector can also refer to a quantity that rotates in an analogous way under isospin or other internal symmetries.

VEV: Vacuum expectation value. The expectation value of a spin-0 field or operator in the lowest energy (vacuum) state of a field theory. If nonzero, it may imply SSB.

Virtual particle: A particle that does not obey the classical relation $E^2 = \vec{p}^2 + m^2$ between its energy, momentum, and mass. It cannot exist as a real particle, but is allowed a fleeting existence due to the uncertainty relations in quantum mechanics, during which it can travel a distance of $\mathcal{O}(1/m)$.

$W^{\pm}$, Z: The heavy-gauge bosons that mediate the weak interactions.

WCC: Weak charged current. Weak charge-changing transitions such as $\nu_\mu \to \mu^- W^+$ in which an electrically charged W boson is emitted or absorbed.

Weak angle (θ_W): The angle $\theta_W \equiv \tan^{-1} g'/g$, where g and g' are the $SU(2)$ and $U(1)$ gauge couplings. θ_W is ubiquitous in the electroweak theory, especially in the Z properties, and has the value $\sin^2 \theta_W \sim 0.23$.

Weak eigenstate: A particle in an irreducible representation of $SU(2) \times U(1)$. It may be a quantum superposition of mass eigenstates.

Weak interaction: The weak short-range interaction responsible for, e.g., β decay, neutrino scattering, and atomic parity violation.

Weak universality: The statement that the transition strength for every $SU(2)$ doublet is the same except for the effects of the unitary CKM and PMNS matrices. Universality is a necessary consequence of $SU(2)$ gauge invariance.

Width (Γ): The uncertainty in the mass of a particle due to its finite lifetime $\tau = \Gamma^{-1}$.

WIMP: Weakly interacting massive particle. A very massive stable or quasi-stable neutral particle with weak interactions, such as the neutralinos in supersymmetry. Such a particle would typically yield a dark matter density in the ballpark of what is observed (the WIMP miracle).

WNC: Weak neutral current. Diagonal weak transitions such as $\nu_\mu \rightarrow \nu_\mu Z$ in which the neutral Z boson is emitted or absorbed.

Yang-Mills theory: A nonabelian gauge theory generalizing QED.

Yukawa interaction: Originally, the description of the strong interactions by the exchange of spin-0 pions. The term is now used for any interaction between spin-0 particles and fermions.

BIBLIOGRAPHY

Pedagogical introductions and reviews are indicated by a double dagger ($\ddagger$) preceding the title.

Aad, G., et al. (2015). "Combined Measurement of the Higgs Boson Mass in pp Collisions at $\sqrt{s} = 7$ and 8 TeV with the ATLAS and CMS Experiments." *Phys. Rev. Lett. 114*, 191803.

Abbott, B. P., et al. (2016). "Observation of Gravitational Waves from a Binary Black Hole Merger." *Phys. Rev. Lett. 116*(6), 061102.

Ade, P.A.R., et al. (2015). "Planck 2015 Results. XIII. Cosmological Parameters." arXiv:1502.01589 [astro-ph.CO].

Adelberger, E. G., J. H. Gundlach, B. R. Heckel, S. Hoedl, and S. Schlamminger (2009). $\ddagger$"Torsion Balance Experiments: A Low-energy Frontier of Particle Physics." *Prog. Part. Nucl. Phys. 62*, 102–34.

Adler, S. L. (2014). "Where Is Quantum Theory Headed?" *J. Phys. Conf. Ser. 504*, 012002.

Aoyama, T., M. Hayakawa, T. Kinoshita, and M. Nio (2015). "Tenth-Order Electron Anomalous Magnetic Moment— Contribution of Diagrams without Closed Lepton Loops." *Phys. Rev. D91*(3), 033006.

Arkani-Hamed, N., S. Dimopoulos, and G. R. Dvali (1998). "The Hierarchy Problem and New Dimensions at a Millimeter." *Phys. Lett. B429*, 263–72.

Arkani-Hamed, N., and J. Maldacena (2015). "Cosmological Collider Physics." arXiv:1503.08043 [hep-th].

Arkani-Hamed, N., and J. Trnka (2014). "The Amplituhedron." *JHEP 10*, 1–33.

Baer, H. W., and X. Tata (2006). *Weak Scale Supersymmetry: From Superfields to Scattering Events*. Cambridge: Cambridge University Press.

Bahcall, J. N. (1989). ‡*Neutrino Astrophysics*. Cambridge: Cambridge University Press.

Barger, V., D. Marfatia, and K. Whisnant (2012). ‡*The Physics of Neutrinos*. Princeton, NJ: Princeton University Press.

Barger, V. D., and R. Phillips (1997). ‡*Collider Physics*. 2nd ed. Frontiers in Physics. Redwood City, CA: Westview Press.

Barnes, K. J. (2010). ‡*Group Theory for the Standard Model of Particle Physics and Beyond*. Boca Raton, FL: CRC Press.

Berenstein, D. (2014). "TeV-Scale Strings." *Ann. Rev. Nucl. Part. Sci. 64*, 197–219.

Bouchendira, R., P. Clade, S. Guellati-Khelifa, F. Nez, and F. Biraben (2011). "New Determination of the Fine Structure Constant and Test of the Quantum Electrodynamics." *Phys. Rev. Lett. 106*, 080801.

Cabibbo, N. (1963). "Unitary Symmetry and Leptonic Decays." *Phys. Rev. Lett. 10*, 531–33.

Canetti, L., M. Drewes, and M. Shaposhnikov (2012). ‡"Matter and Antimatter in the Universe." *New J. Phys. 14*, 095012.

Carlson, C. E. (2015). ‡"The Proton Radius Puzzle." *Prog. Part. Nucl. Phys. 82*, 59–77.

Charles, J., et al. (2005). "CP Violation and the CKM Matrix: Assessing the Impact of the Asymmetric *B* Factories." *Eur. Phys. J. C41*, 1–131.

Christenson, J., J. Cronin, V. Fitch, and R. Turlay (1964). "Evidence for the 2π Decay of the K_2^0 Meson." *Phys. Rev. Lett. 13*, 138–40.

Commins, E. D., and P. H. Bucksbaum (1983). ‡*Weak Interactions of Leptons and Quarks.* Cambridge: Cambridge University Press.

Damour, T., and J. F. Donoghue (2008). "Constraints on the Variability of Quark Masses from Nuclear Binding." *Phys. Rev. D78*, 014014.

Dine, M. (2015). *Supersymmetry and String Theory: beyond the Standard Model.* 2nd ed. Cambridge: Cambridge University Press.

Eichten, E., K. D. Lane, and M. E. Peskin (1983). "New Tests for Quark and Lepton Substructure." *Phys. Rev. Lett. 50*, 811–14.

Fritzsch, H. (2014). ‡"Quarks and QCD." *Int. J. Mod. Phys. A29*(15), 1430038.

Gaillard, M., and B. W. Lee (1974). "Rare Decay Modes of the K-Mesons in Gauge Theories." *Phys. Rev. D10*, 897–916.

Gaisser, T., and F. Halzen (2014). ‡"IceCube." *Ann. Rev. Nucl. Part. Sci. 64*, 101–23.

Gelmini, G. B. (2015). ‡"TASI 2014 Lectures: The Hunt for Dark Matter." arXiv:1502.01320 [hep-ph].

Georgi, H., and S. L. Glashow (1974). "Unity of All Elementary Particle Forces." *Phys. Rev. Lett. 32*, 438–41.

Georgi, H., H. R. Quinn, and S. Weinberg (1974). "Hierarchy of Interactions in Unified Gauge Theories." *Phys. Rev. Lett. 33*, 451–54.

Giudice, G. F. (2008). ‡"Naturally Speaking: The Naturalness Criterion and Physics at the LHC." arXiv:0801.2562 [hep-ph].

Glashow, S. L. (1980). ‡"Towards a Unified Theory: Threads in a Tapestry." *Rev. Mod. Phys. 52*, 539–43.

Glashow, S. L., J. Iliopoulos, and L. Maiani (1970). "Weak Interactions with Lepton-Hadron Symmetry." *Phys. Rev. D2*, 1285–92.

Graham, P. W., I. G. Irastorza, S. K. Lamoreaux, A. Lindner, and K. A. van Bibber (2015). ‡"Experimental Searches for the Axion and Axion-like Particles." *Ann. Rev. Nucl. Part. Sci. 65*(1), 485–514.

Green, A. M. (2015). "Primordial Black Holes: Sirens of the Early Universe." *Fundam. Theor. Phys. 178*, 129–49.

Griffiths, D. J. (2008). ‡*Introduction to Elementary Particles.* 2nd rev. version. New York: Wiley.

Gross, D. J. (2005). ‡"The Discovery of Asymptotic Freedom and the Emergence of QCD." *Rev. Mod. Phys. 77*, 837–49.

Gross, D. J., and F. Wilczek (1973). "Ultraviolet Behavior of Nonabelian Gauge Theories." *Phys. Rev. Lett. 30*, 1343–46.

Guth, A. H. (1981). "The Inflationary Universe: A Possible Solution to the Horizon and Flatness Problems." *Phys. Rev. D23*, 347–56.

Ibrahim, T., and P. Nath (2008). "CP Violation from Standard Model to Strings." *Rev. Mod. Phys. 80*, 577–631.

Karshenboim, S. G. (2005). "Precision Physics of Simple Atoms: QED Tests, Nuclear Structure and Fundamental Constants." *Phys. Rept. 422*, 1–63.

Kinoshita, T. (1990). *Quantum Electrodynamics.* Singapore: World Scientific.

Kleinknecht, K. (2003). ‡*Uncovering CP Violation: Experimental Clarification in the Neutral K Meson and B Meson Systems.* Springer Tracts in Modern Physics. Berlin: Springer.

Kobayashi, M., and T. Maskawa (1973). "CP Violation in the Renormalizable Theory of Weak Interaction." *Prog. Theor. Phys. 49*, 652–57.

Kolb, E. W., and M. S. Turner (1990). ‡*The Early Universe.* Redwood City, CA: Addison-Wesley.

Langacker, P. (2010). ‡*The Standard Model and Beyond.* Boca Raton, FL: Taylor and Francis.

Langacker, P. (2012). ‡"Grand Unification." *Scholarpedia.* www.scholarpedia.org/article/Grand_unification.

Lederman, L. M. (1963). ‡"The Two-Neutrino Experiment." *Sci. Am. 208N3*, 60–70.

Leutwyler, H. (2014). ‡"On the History of the Strong Interaction." *Mod. Phys. Lett. A29*, 1430023.

Linde, A. (2015). ‡"A Brief History of the Multiverse." arXiv:1512.01230 [hep-th].

Loeb, A. (2010). ‡*How Did the First Stars and Galaxies Form?* Princeton, NJ: Princeton University Press.

Maldacena, J. M. (1999). "The Large N Limit of Superconformal Field Theories and Supergravity." *Int. J. Theor. Phys. 38*, 1113–33. [*Adv. Theor. Math. Phys. 2*, 231 (1998)].

Mann, R. (2010). ‡*An Introduction to Particle Physics and the Standard Model.* Boca Raton, FL: CRC Press.

Mohr, P. J., B. N. Taylor, and D. B. Newell (2012). "CODATA Recommended Values of the Fundamental Physical Constants: 2010." *Rev. Mod. Phys. 84*, 1527–1605.

Murray, W., and V. Sharma (2015). ‡"Properties of the Higgs Boson Discovered at the Large Hadron Collider." *Ann. Rev. Nucl. Part. Sci. 65*(1), 515–54.

Nath, P., and P. Fileviez Perez (2007). ‡"Proton Stability in Grand Unified Theories, in Strings and in Branes." *Phys. Rept. 441*, 191–317.

Olive, K. A., et al. (2014). "Review of Particle Physics (RPP)." *Chin. Phys. C38*, 090001. http://pdg.lbl.gov.

Pais, A. (1986). ‡*Inward Bound: Of Matter and Forces in the Physical World.* Oxford: Clarendon Press.

Peccei, R., and H. R. Quinn (1977). "CP Conservation in the Presence of Instantons." *Phys. Rev. Lett. 38*, 1440–43.

Politzer, H. D. (1973). "Reliable Perturbative Results for Strong Interactions?" *Phys. Rev. Lett. 30*, 1346–49.

Quigg, C. (2013). ‡*Gauge Theories of the Strong, Weak, and Electromagnetic Interactions.* 2nd ed. Princeton, NJ: Princeton University Press.

Quigg, C. (2015). ‡"Electroweak Symmetry Breaking in Historical Perspective." *Ann. Rev. Nucl. Part. Sci. 65*(1), 25–42.

Raby, S. (2009). "SUSY GUT Model Building." *Eur. Phys. J. C59*, 223–47.

Randall, L., and R. Sundrum (1999). "A Large Mass Hierarchy from a Small Extra Dimension." *Phys. Rev. Lett. 83*, 3370–73.

Sachdev, S. (2013). ‡"Strange and Stringy." *Sci. Am. 308*, 44–51.

Sakharov, A. D. (1967). "Violation of CP Invariance, C Asymmetry, and Baryon Asymmetry of the Universe." *JETP Lett. 5*, 24–27.

Salam, A. (1968). "Weak and Electromagnetic Interactions." In *Elementary Particle Theory*, ed. N. Svartholm. Stockholm: Almquist and Wiksells, 1968, 367–77.

Salam, A. (1980). ‡"Gauge Unification of Fundamental Forces." *Rev. Mod. Phys. 52*, 525–38 [*Science 210*, 723 (1980)].

Schael, S., et al. (2013). "Electroweak Measurements in Electron-Positron Collisions at W-Boson-Pair Energies at LEP." *Phys. Rept. 532*, 119–244.

Schellekens, A. (2013). ‡"Life at the Interface of Particle Physics and String Theory." *Rev. Mod. Phys. 85*(4), 1491–1540.

Schwinger, J. S. (1958). *Selected Papers on Quantum Electrodynamics.* New York: Dover.

Seiberg, N. (2006). "Emergent Spacetime." arXiv:hep-th/0601234 [hep-th].

Sirlin, A., and A. Ferroglia (2013). ‡"Radiative Corrections in Precision Electroweak Physics: A Historical Perspective." *Rev. Mod. Phys. 85*(1), 263–97.

Streater, R. F., and A. S. Wightman (2000). *PCT, Spin and Statistics, and All That*. Princeton, NJ: Princeton University Press.

Thomson, M. (2013). ‡*Modern Particle Physics*. Cambridge: Cambridge University Press.

Tully, C. G. (2011). ‡*Elementary Particle Physics in a Nutshell*. Princeton, NJ: Princeton University Press.

Uzan, J.-P. (2003). ‡"The Fundamental Constants and Their Variation: Observational Status and Theoretical Motivations." *Rev. Mod. Phys. 75*, 403–55.

Weinberg, S. (1967). "A Model of Leptons." *Phys. Rev. Lett. 19*, 1264–66.

Weinberg, S. (1980a). ‡"Conceptual Foundations of the Unified Theory of Weak and Electromagnetic Interactions." *Rev. Mod. Phys. 52*, 515–23. [*Science 210*, 1212 (1980)].

Weinberg, S. (1980b). "Varieties of Baryon and Lepton Nonconservation." *Phys. Rev. D22*, 1694–700.

Weinberg, S. (1983). ‡*The First Three Minutes: A Modern View of the Origin of the Universe*. London: Fontana.

Weinberg, S. (1989). ‡"The Cosmological Constant Problem." *Rev. Mod. Phys. 61*, 1–23.

Weinberg, S. (1995). *The Quantum Theory of Fields*, vol I. Cambridge: Cambridge University Press.

Weinberg, S. (2008). *Cosmology*. Oxford: Oxford University Press.

Weinberg, S. (2015). ‡*To Explain the World: The Discovery of Modern Science*. New York: Harper.

Yang, C.-N., and R. L. Mills (1954). "Conservation of Isotopic Spin and Isotopic Gauge Invariance." *Phys. Rev. 96*, 191–95.

Yoshimura, M. (1978). "Unified Gauge Theories and the Baryon Number of the Universe." *Phys. Rev. Lett. 41*, 281–84. [Erratum: *Phys. Rev. Lett. 42*, 746 (1979)].

Zwiebach, B. (2009). *A First Course in String Theory*. 2nd ed. Cambridge: Cambridge University Press.

INDEX